Thermoluminescence Dosimetry

Medical Physics Handbooks
Other books in the series

1 **Ultrasonics**
J P Woodcock

2 **Computing Principles and Techniques**
B L Vickery

3 **Physical Principles of Audiology**
P M Haughton

4 **Urodynamics**
D J Griffiths

6 **Fundamentals of Radiation Dosimetry**
J R Greening

7 **Radiotherapy Treatment Planning**
R F Mould

8 **Nuclear Particles in Cancer Treatment**
J F Fowler

Series Editor: **Professor J M A Lenihan**
Department of Clinical Physics and Bio-Engineering
West of Scotland Health Boards, Glasgow

Medical Physics Handbooks 5

Thermoluminescence Dosimetry

A F McKinlay
National Radiological Protection Board,
Harwell

Adam Hilger Ltd, Bristol
in collaboration with the
Hospital Physicists' Association

British Library Cataloguing in Publication Data

McKinlay, A F
Thermoluminescence dosimetry.—
(Medical Physics Handbook; 5 ISSN 0143-0203).
1. Radiology, Medical
2. Thermoluminescence dosimetry
I. Title II. Series
615′.842 R895

ISBN 0-85274-520-6

Published by Adam Hilger Ltd,
Techno House, Redcliffe Way, Bristol BS1 6NX

The Adam Hilger book-publishing imprint is owned by The Institute of Physics

Filmset and printed in Great Britain by Page Brothers (Norwich) Ltd, Mile Cross Lane, Norwich NR6 6SA.

To Carolyne, Alison and Fiona

Contents

Preface

Occupational and medical exposure of individuals to ionising radiations is a common feature of present day life. The detrimental effects of excessive exposure are well established. In medicine, ionising radiations are used in the diagnosis and treatment of disease. Consequently the accurate measurement of exposure and dose is important in the monitoring of persons who are occupationally exposed to ionising radiations, and in medicine to achieve the minimum exposures consistent with good diagnosis and treatment. Such measurements have traditionally been carried out using devices such as air ionisation chambers and film badges, although thermoluminescence dosemeters (TLDs) are increasingly being used.

The purpose of this text is to provide an introduction to the use of TLDs in ionising radiation dose measurement with particular emphasis on clinical dosimetry. Although written for graduate scientists, it is also suitable for non-graduate technical staff concerned with the routine use of TLDs.

In a text of this length it is impossible to provide in-depth information on all aspects of thermoluminescence dosimetry (TLD), and in view of the vast amount of data on the subject in the scientific literature the choice of examples and data for inclusion has been difficult, and by necessity very selective. Nevertheless, the text contains over 170 references selected to represent the scope of application of TLDs and the many scientists who have made important contributions to the subject.

I have written the text under three main themes. The first, comprising Chapters 1, 2 and 3, provides a progressive introduction to the subject via discussions on luminescence phenomena, theoretical aspects of thermoluminescence including glow curve theory, proposed trapping and recombination models in lithium fluoride, supralinearity, phototransferred thermoluminescence, and the thermoluminescence (TL) properties and dose measurement characteristics of a number of commonly used phosphors.

Secondly, Chapters 4 and 5 describe the use of TLDs for specific measurement applications including clinical, personal, environmental, charged particle, neutron, and mixed-field dosimetry. Thermoluminescence dating techniques are also described.

Thirdly, the principles of design and operation of TLD readers are described in Chapter 6. Chapter 7 deals with the practical problems and potential pitfalls in the use of TLDs, including those associated with annealing, storage and handling, irradiation, calibration and readout.

I should like to express my thanks to those colleagues at the National Radiological Protection Board, Harwell, whose constructive criticisms have contributed greatly to the preparation of this text—in particular my thanks to Dr J A Dennis, Dr S Rae, Dr J A Reissland, Mr F Harlen, Dr C M H Driscoll and Ms I Clark, NRPB Glasgow. I am also indebted to Mr J Millham and his photographic section, to the library staff, and to Mrs V Goodwin and her typing section for their patient help in the preparation of the manuscript. Whilst thanking these individuals, I must point out that the text in whole or in part does not necessarily reflect the opinions or policy of the National Radiological Protection Board.

Alastair McKinlay

1 Luminescence Phenomena

1.1 Introduction

Luminescence describes the process of emission of optical radiation from a material from causes other than heating it to incandescence. Luminescent materials (luminophors—light bearers) can absorb energy, store a fraction of it, and convert it into optical radiation which is then emitted. Luminescence embraces many similar but specific effects, each of which can be described by the addition of a prefix to the term luminescence. The prefix generally, but not invariably, describes the means by which the material receives its excitation energy. A number of examples of this nomenclature are shown in table 1.1. Thermo-

Table 1.1 Luminescence effects.

Luminescence effect	Means of excitation
Photoluminescence	Optical photons (ultraviolet, visible, infrared)
Triboluminescence	Rubbing or grinding
Chemiluminescence	Chemical energy
Bioluminescence	Biochemical energy
Cathodoluminescence	Cathode rays
Electroluminescence	Electric field
Radioluminescence	Ionising radiation
Sonoluminescence	Sound waves
Fluorescence (prompt emission)	Various
Phosphorescence (delayed emission)	Various
Thermoluminescence (thermally accelerated emission)	Various

luminescence (TL), fluorescence and phosphorescence are particular forms of luminescence, related not to the means of excitation (which for all three may be varied), but to the time scale over which the emission of luminescence takes place.

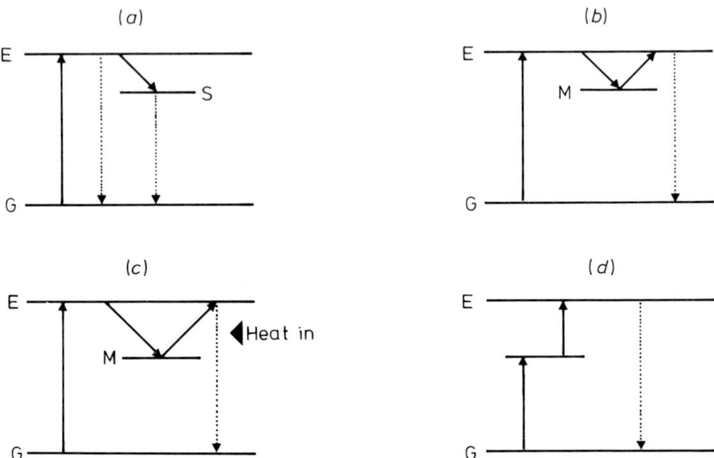

Figure 1.1 Simple examples of luminescence processes. (*a*) Fluorescence: the prompt return of an electron from an excited state either directly to the ground state or via an allowed transition from an intermediate state S (relaxation). Where the excitation energy is provided by photons, the luminescence emission photons are of longer wavelength (Stokes' law). (*b*) Phosphorescence: return of an electron from an excited state to the ground state is delayed by the metastable state M. Direct transition of an electron from the metastable to the ground state is forbidden. (*c*) Thermoluminescence: the return of electrons trapped in the metastable state is speeded up by heating the material. (*d*) Anti-Stokes' luminescence: the excitation of an electron is achieved by the absorption of two or more photons.

Luminescence excitation involves the transfer of energy to electrons and their displacement to a higher energy state (excited state E) as illustrated in figure 1.1. Should the electrons return promptly to their original energy state (ground state G) with the emission of optical radiation as shown in figure 1.1(*a*), then the process is called fluorescence.

However, if, due to the presence of an electron trap (metastable state M), the return of electrons to the ground state is delayed, the process is termed phosphorescence, as shown in figure 1.1(*b*). The transition of electrons directly from a metastable state to the ground state is forbidden. The metastable state represents a shallow electron trap and electrons returning from it to the excited state require energy. This energy can be supplied in the form of optical radiation (photostimulation) or as heat (thermal stimulation). The probability p of escape of an electron from a metastable state to an excited state is governed by the Boltzmann equation.

$$p = s \exp(-\Delta E/kT),$$

where s is a constant, ΔE is the energy difference between states E and M (commonly called the trap depth), k is Boltzmann's constant, and T is the temperature in kelvin.

By raising the temperature, the probability of escape of an electron is increased, and this effectively accelerates the phosphorescence process as progressively deeper metastable states empty with increasing temperature. This process is called thermoluminescence (illustrated in figure 1.1*c*), and is quite different from incandescence, in which the application of heat causes vigorous vibrations, collisions and excitation of all the atoms of a material, resulting in an emission spectrum approximating to that of a black-body radiator at the same temperature. The form of this emission curve is determined by temperature according to Planck's law, as illustrated in figure 1.2. In contrast, the thermoluminescence emission spectrum of a material depends on the species of luminescent atoms present. While recognising that there are many possible sources of excitation which will produce thermoluminescence, such as some of those listed in table 1.1, the term thermoluminescence will be used throughout this text to describe the thermoluminescence produced as a result of the absorption of ionising radiation, in accordance with common practice.

It is not always clear whether a particular emission is fluorescence or phosphorescence. If, after removal of excitation, the decay of the excited atom is of the order of 10 ns or so, the process is undoubtedly fluorescence, whereas should the emission persist for a period longer than 100 ms (which the eye will resolve) it is undoubtedly phosphorescence. For intermediate decay periods it may be necessary to examine the effect of raising the temperature of the material to decide whether the emission is fluorescence or phosphorescence. Fluorescence is essentially independent of temperature, whereas, as we have seen, phosphorescence is accelerated by a rise in temperature according to Boltzmann's equation.

For completeness it is worth noting one other form of luminescence transition: anti-Stokes' luminescence. In this process the wavelength of the luminescence emission is shorter than that of the excitation radiation, in violation of Stokes' law. Conservation of energy is maintained, however, as two or more excitation photons are necessary for each luminescence photon emitted, as illustrated in figure 1.1(*d*). Materials displaying anti-Stokes' luminescence are used in the production of certain types of semiconductor photoemissive device.

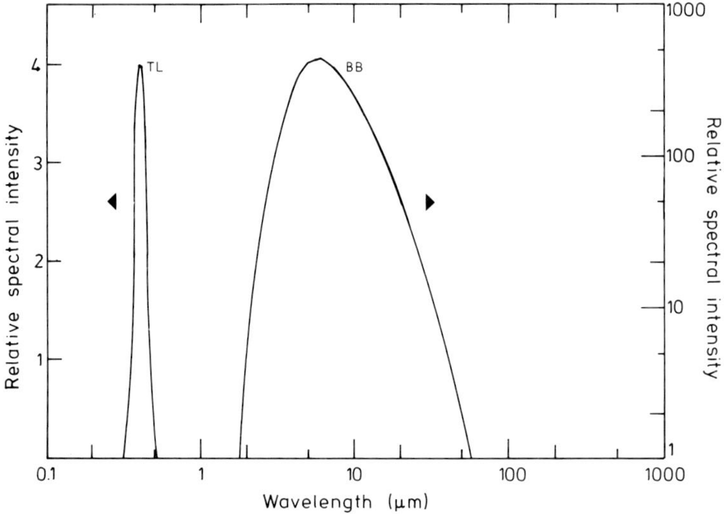

Figure 1.2 Thermoluminescence (TL) and black-body (BB) emission from LiF heated to 240 °C.

1.2 Historical Background

The emission of light from materials has long been a subject of interest to man. The glow from micro-organisms, fungi, insects and heated minerals were probably the first luminescence phenomena observed. The classical literature contains many references to luminescence phenomena, many of which were associated with the supernatural, and were incorporated into myths and legends.

Great advances were made in the logical investigation of luminescence in the seventeenth century. At this time science was beginning to be separated from magic, and the emission of light from liquids and solids became a topic of great interest, speculation and debate. Scientific works of the seventeenth century contain many dissertations and observations on luminescence phenomena; the ability of phosphorus (first isolated and appropriately named in 1669) to glow in the dark was of particular interest. This is a phenomenon which we now know to be a form of chemiluminescence caused by oxidation. The subject of a letter written to the Royal Society in 1677 was 'a factitious [artificial] strong matter or paste shining in the dark like a glowing coal, after it has been a little while exposed to the day or candle light.'

Robert Boyle (1663) presented some observations on the luminescence behaviour of a diamond which belonged to a Mr Clayton. Detailing a number of experiments (all of which he is careful to point out, of a non-destructive nature) to discover the nature and cause of the luminescence, he wrote†:

> Eleventhly, I alſo brought it to ſome kind of Glimmering Light, by taking it into Bed with me, and holding it a good while upon a warm part of my Naked Body.
>
> Twelfthly, To ſatisfie my ſelf, whether the Motion introduc'd into the Stone did generate the Light upon the account of its producing Heat there, I held it near the Flame of a Candle, till it was qualify'd to ſhine pretty well in the Dark,

In the late nineteenth century there was much interest in the properties of ultraviolet radiation, and in x-rays, after their discovery in 1895. It was found that these radiations had the ability to produce luminescence in certain minerals, particularly those containing manganese as an impurity. Among these were calcium fluoride, calcium sulphate and zinc orthosilicate (willemite).

The association of luminescence, and in particular thermoluminescence, with exposure of a material to the radiations emitted by radioactive salts was observed by the pioneers of research in radioactivity. Both

† Reproduced by courtesy of the Royal Society of London.

fluorite and calcium fluoride containing manganese were used to detect ionising radiations. Marie Curie observed and noted the thermoluminescence of calcium fluoride exposed to radium, and in her doctoral thesis (1904) she wrote:

> *Production of Thermo-luminosity.* Certain bodies such as fluorite, become luminous when heated; they are thermo-luminescent. Their luminosity disappears after some time, but the capacity of becoming luminous afresh through heat is restored to them by the action of a spark, and also by the action of radium. Radium can thus restore to these bodies their thermo-luminescent property.

By the 1940s, a reasonably clear understanding of the general processes involved in thermoluminescence had been developed. It had been noted that luminescence and photoconductivity were more apparent in those crystals which had undergone thermal treatment, which would cause disorder and stress, or when it was known that foreign atoms were present in the crystal lattice. We shall return to this topic in Chapter 2. The optical properties and photoconductivity of what were regarded as 'pure' alkali halides and ones containing an excess of alkali metal halide ions were examined, and it was postulated that negative ion vacancies could form electron traps. It was also postulated that luminescence in crystals containing impurities was characteristic of the foreign impurity atoms and not of the host lattice. These foreign atoms were given the name 'luminescent centres'. Randall and Wilkins (1945) related the fundamental similarity between phosphorescence and thermoluminescence and explained both using a model of electron traps in the crystal lattice. They also formulated the theory of the thermoluminescence glow curve which is discussed in detail in Chapter 2.

1.3 Development of Thermoluminescent Phosphors

In the 1930s and 1940s the scientific investigation of luminescence ran parallel with its commercial and industrial applications. It was used in paints, cathode-ray tubes and fluorescent lights. The original commercial fluorescent lights used willemite and zinc beryllium silicate.

In the late 1940s, Daniels and his co-workers were the first to use thermoluminescence to make quantitative measurements of radiation exposure. Their main initial interests lay in the thermoluminescence of geological specimens, and the study of the TL properties of alkali halides. They examined the glow peak structure of alkali halides and their isothermal fading, and applied them to the measurement of ionising radiation exposure. They concluded that lithium fluoride (LiF—obtained from the Harshaw Chemical Company) in the form of small single crystals was most suitable. Such crystals were also used at that time for the measurement of internal radiation exposure to patients receiving radiation therapy. However, it was found that there were unacceptably large variations in thermoluminescence sensitivity of crystals within each batch, and that these variations could be reduced considerably by grinding the crystals to a powder which was then mixed and used in pellet form. However, the grinding introduced a spurious thermoluminescence effect, commonly called tribothermoluminescence (TTL), so that other materials, including alumina (Al_2O_3), were preferred.

At that time also, photostimulated radioluminescence was being studied with a view to its use as a practical dosemeter. This is the so-called phosphate glass method of dosimetry. These dosemeters were produced in great numbers, primarily for use by the US armed forces. They were also produced in the form of needles, and found applications in clinical dosimetry.

Calcium sulphate activated with manganese was examined, but displayed a high fading rate. Little further work was done until 1957 when Ginther and Kirk (1957) investigated the properties of calcium fluoride activated with manganese. As detailed in Chapter 3, this phosphor has the advantage of a high TL sensitivity, although the response is markedly sensitive to photons at low energies. An early form of dosemeter based on this material consisted of a phosphor layer deposited on a graphite plate element enclosed in a bulb filled with a noble gas.

During further studies of LiF in 1960 it was found that the material then being manufactured by Harshaw had a very low TL sensitivity compared with that of Daniels' original material. It was realised that the superior crystal refinement techniques developed in the interim, and then being used, had removed unknown trace elements which were necessary for efficient thermoluminescence. Titanium (Ti) was identified as one such impurity, and Harshaw subsequently incorporated this and other elements in the LiF to produce a phosphor with a high TL sensitivity. This material is the basis of what is now generally regarded as the 'standard' TL phosphor: Harshaw TLD 100.

Thermal annealing and other studies revealed the thermal complexity of the material and how careful one had to be with its thermal treatment in order to achieve reproducible results. The material became available in new physical forms, and was incorporated in polytetrafluoroethylene

(PTFE) matrices as discs, tapes and rods. The advantages of these dosemeters were immediately obvious: they used LiF phosphor, the thermoluminescence characteristics of which were well documented, and the phosphor was in a regular-geometry solid form, so that the main problems associated with the use of loose powder did not arise, i.e. tribothermoluminescence and the weighing or volume dispensing of powder. The useful thermoluminescence properties of these discs and rods were essentially identical to those of loose powder, but they could be conveniently handled, washed, annealed and sterilised, and so presented new possibilities for use as personal and clinical dosemeters. A further major development was the production of the extruded-ribbon form of dosemeter—the so-called chip. This was originally produced only from lithium fluoride, but other phosphors are now available in this form. Because extruded ribbons contain no non-thermoluminescent binding material, their TL sensitivity per unit weight is higher than for PTFE-based dosemeters. Also they can be annealed at the more desirable higher temperature of 400 °C which the PTFE-based dosemeter cannot withstand.

The complex annealing requirements, and the enhancement of TL response at low photon energies when tissue equivalence is required, are the main disadvantages of lithium fluoride. In 1965 lithium borate doped with manganese ($Li_2B_4O_7$: Mn) was introduced as a tissue-equivalent phosphor with a simple annealing procedure, although this material has a TL spectral emission peak at 600 nm in the red, compared with the blue (400 nm) emission of LiF. Consequently, since almost all commercially available TL detection systems have a blue-sensitive photomultiplier, the sensitivity of lithium borate is generally less than that of lithium fluoride. It has also proved difficult to produce regular-geometry solid forms of dosemeter from lithium borate. For these reasons it has not generally replaced lithium fluoride, except for special applications where its superior tissue equivalence of atomic number and electron density is a great advantage in, for example, certain clinical and biological dosimetry applications.

Over the last 15 years the investigation of thermoluminescence and other related luminescence dosimetry techniques has expanded enormously, and up to 1979 there have been five international conferences on luminescence dosimetry. Many phosphor materials have been studied, become commercially available, and have been applied to many areas of ionising radiation dosimetry, including personal, environmental, clinical, and charged particle and neutron dosimetry.

Thermoluminescence has also been applied to many other areas, including the authentication of works of art, dating of pottery and other materials, examination of lunar soil, and analysis of the thermal history of meteorites. As manufacturing interest in the production of TL phosphors has grown steadily, so too has the number of commercially available instruments for reading the TL signal. Manual readers can be obtained for routine clinical, personal and environmental dosimetry studies. With many readers the addition of various accessories and optional extras enables glow curve analysis, material studies, etc, to be carried out. Fully and semi-automated readout systems have been developed for large-scale personal dosimetry applications.

2 Theoretical Aspects of Thermoluminescence

2.1 Introduction

The mechanism of thermoluminescence is complex, and although general theoretical models can be postulated, difficulties arise when specific dosimetric materials are considered—particularly the common phosphor lithium fluoride doped with magnesium and titanium—LiF : Mg : Ti. Several theoretical models for the thermoluminescence process in this material have been suggested but none fully explains all of the experimentally observed phenomena, particularly the precise roles of magnesium, titanium and hydroxyl (OH) ions in the trapping and luminescence recombination process (Niewiadomski 1976). The models have been developed on the basis of experimental evidence but have proved singularly unsuccessful in the prediction of experimental behaviour of materials which might lead to a greatly improved phosphor. It is sobering to note that the adopted 'standard' LiF : Mg : Ti phosphor, Harshaw TLD 100 and its isotopic variants, were the end products of what was essentially a chance discovery, and despite many studies of this material it has so far proved impossible to produce a phosphor which is a significant improvement.

One of the major problems with models has been that many of the experimental observations on which they have been based can be observed only at very high levels of absorbed dose, such as in optical density studies. It is difficult to assess how far one can relate these effects to the prediction of effects at the normally low levels of absorbed dose encountered in the everyday use of a phosphor. Each TL phosphor is unique, displaying very different characteristics. The general models which have been developed provide no basis for the prediction of these specific characteristics.

A general theoretical mechanism for thermoluminescence may be developed by referring to the simplest of all multi-atomic crystalline structures—the alkali halides. They consist of two interpenetrating cubic lattices of alkali and halogen ions, as illustrated in figure 2.1(*a*). The structure shown represents that which would exist in the ideal case of a perfect crystal. All real crystals contain lattice defects of various kinds and these play an all-important role in the thermoluminescence process.

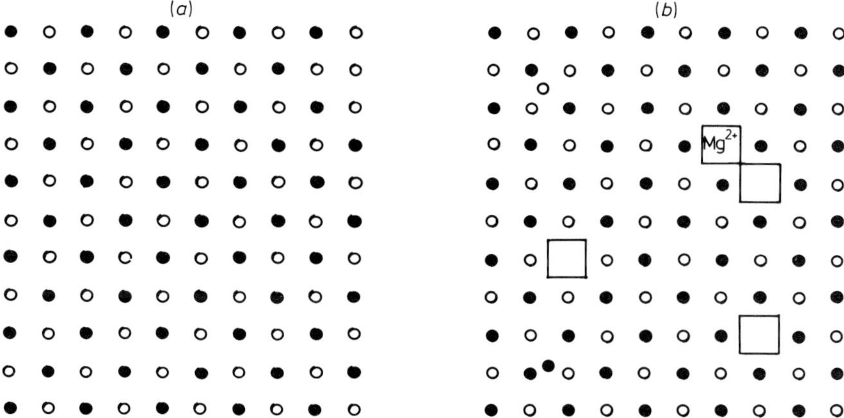

Figure 2.1 Ionic structure of (*a*) an 'ideal' perfect alkali halide crystal, and (*b*) a 'real' imperfect crystal containing defects of various types. Also shown is a divalent magnesium ion–alkali metal ion vacancy dipole. ● halogen ion, ○ alkali metal ion.

There are three categories of lattice defects:

(i) thermal or intrinsic defects;
(ii) extrinsic defects or substitutional impurity ions; and
(iii) radiation-induced defects.

For intrinsic defects, the higher the temperature of the lattice the greater is the number of defects, a point which is important when one considers the thermal treatment of thermoluminescence phosphors. If a phosphor is quenched rapidly from a high to a low temperature then it is possible effectively to 'freeze in' the number of defects appropriate to the higher temperature.

At any given temperature the lower the energy for defect formation the greater is the probability of the formation of such a defect. Simple defects are much more favourable energetically than large complex ones.

2.2 The Role of Lattice Defects in the Thermoluminescence Process

The presence of defects in a material is important for the thermoluminescence process. For the moment we will consider only the role of intrinsic defects in the electron trapping process.

A negative ion vacancy, illustrated in figure 2.1(*b*), is essentially a region of excess positive charge and as such may be regarded as a potential trap for a free electron. Electron capture creates an F centre. Similarly, a region of excess negative charge will be a potential trap for free positive charges (holes). The analogue 'anti-centre' of an F centre would be a hole trapped at a positive ion vacancy (a V_F centre), but there is doubt as to whether this particular centre generally exists (Pick 1972).

The energy band structure for an ideal crystal may be represented by an energy band diagram as shown in figure 2.2(*a*). The valence band is representative of all electrons held in bound states, and the conduction band is representative of all electrons in unbound states which are free to migrate through the crystal lattice. In the case of the ideal electrically insulating crystal under discussion the conduction band will be empty and all electrons will reside in the valence band. The conduction band and the valence band are widely separated in energy by the so-called 'forbidden gap.' Without the influence of external forces it is highly improbable for an electron to be able to traverse the forbidden gap from

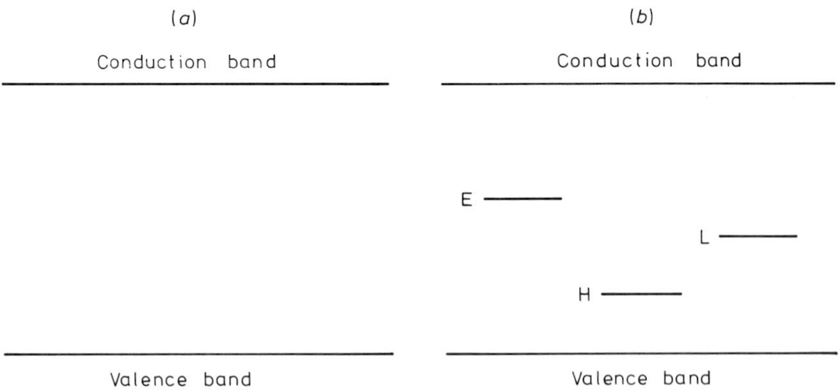

Figure 2.2 Energy band diagram of (*a*) an 'ideal' electrically insulated crystal, and (*b*) a 'real' crystal containing defects giving rise to various centres.

the valence to conduction band. However, in the case of a real disordered crystal containing defects of a simple and/or complex nature, other allowed energy levels exist in the forbidden gap region, as illustrated in figure 2.2(*b*).

In the description of the general model which follows we shall suppose that the energy level labelled E represents an electron trap and that level H represents a hole trap. L is a luminescence centre where electrons and holes may recombine with photon emission. We shall not concern ourselves with the precise nature of the structure of the traps at this stage. The glow curve theory of Randall and Wilkins (1945) is used and assumes that no retrapping of released electrons occurs (first-order kinetics). The case of equal retrapping and recombination probabilities is called second-order kinetics. Indeed the trapping and recombination processes may not be limited to first- or second-order kinetics but may follow a general 'order' (Chen and Winer 1970).

2.3 A General Model for Thermoluminescence

The production of thermoluminescence in a material by exposure to ionising radiation may be divided into two stages:

(i) ionisation and electron trapping, and

(ii) electron and hole recombination with photon emission.

Figure 2.3 illustrates the energy band configuration for each stage.

Ionising radiation is absorbed in the material and free electrons are produced. With respect to the energy band diagram this is equivalent to transferring electrons from the valence band to the conduction band (step 1). These electrons are now free to move through the crystal (step 2), but if trapping levels such as E are present the electrons may become trapped (step 3). The production of free electrons is associated with the production of free positive holes which may also migrate, in energy terms, via the valence band (step 2′). The holes may become trapped (step 3′).

The trapped electron centres produced, such as F centres, are lattice defect centres and as such their properties are primarily determined by the lattice and the defect. Many hole centres are thermally unstable and may decay rapidly at normal room temperature (step 4′).

The trapped electrons will remain in their traps provided that they do not acquire sufficient energy to escape. This will be determined by two main factors: the depth of the traps, and the temperature of the material. If the temperature of the material is raised trapped electrons may acquire sufficient thermal energy and be released (step 4). Released electrons may recombine with holes at luminescence centres such as L, and the excess energy is radiated as visible or ultraviolet photons (step

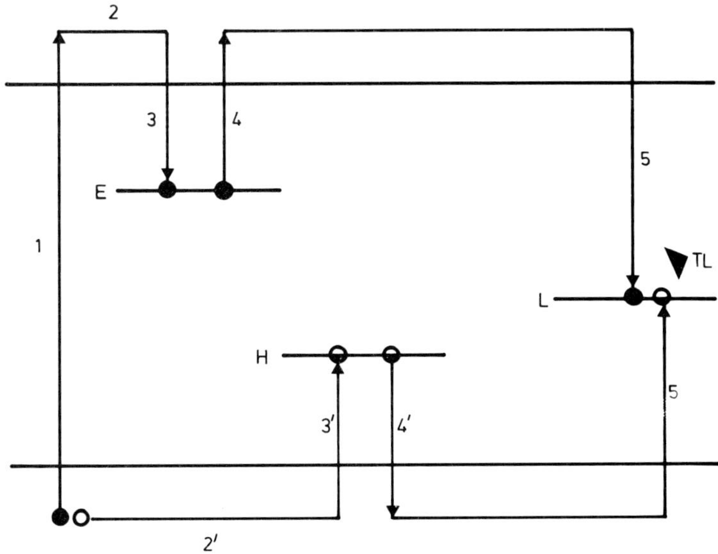

Figure 2.3 A simple energy band model for thermoluminescence. ● electron, ○ hole.

5). While electron capture and delayed recombination with a hole at a luminescence centre is the mechanism of thermoluminescence, other electron–hole recombination processes are possible, i.e. immediate or delayed recombination with subsequent thermal degradation of energy without photon emission, and fluorescence caused by the immediate recombination of holes and electrons at luminescence centres (Attix 1974).

2.4 The Glow Curve

Let us consider a material containing defects which give rise to a single electron trap. The energy depth of the ground state of this trap is, for example, E below the bottom of the conduction band. A trap may also have several excited states. At some time t, the temperature of the material is T (K), and the single electron trap contains n electrons. The energy distribution of electrons within the trap will be described by the Boltzmann distribution, and hence the probability p of release of a single electron from the trap is given by

$$p = s \exp(-E/kT), \tag{2.1}$$

where k is Boltzmann's constant and s is a frequency factor associated with the particular lattice defect. The rate of release of electrons from the trap is

$$-\frac{dn}{dt} = ns \exp(-E/kT). \tag{2.2}$$

Assuming that no electrons released from traps are retrapped, but that all undergo thermoluminescence transitions, the intensity of the thermoluminescence glow I depends on the rate of photon emission and therefore on the rate of release of electrons from traps and their rate of arrival at luminescence centres:

$$I = -C\frac{dn}{dt} = Cns \exp(-E/kT), \tag{2.3}$$

where C is a constant related to luminescence efficiency.

If the material is heated at a uniform rate

$$R = \frac{dT}{dt}, \tag{2.4}$$

then

$$\frac{dn}{dt} = \frac{dn}{dT} \cdot \frac{dT}{dt} = R \cdot \frac{dn}{dT}. \tag{2.5}$$

By substitution in equation (2.2),

$$\frac{dn}{dT} = -\frac{1}{R} ns \exp(-E/kT) \tag{2.6}$$

or

$$\frac{dn}{n} = -\frac{s}{R} \exp(-E/kT)\, dT.$$

Integrating gives

$$\ln\left(\frac{n}{n_0}\right) = -\int_{T_0}^{T} \frac{1}{R} s \exp(-E/kT)\, dT, \tag{2.7}$$

where n_0 is the number of electrons present in the trap at time t_0 and temperature T_0.

Finally, substituting for n in equation (2.3),

$$I = n_0 C \exp - \left[\int_{T_0}^{T} \frac{1}{R} s \exp(-E/kT) \, dT \right] s \exp(-E/kT). \quad (2.8)$$

This is the expression for the glow intensity I from electrons trapped at a single trapping level E. The plot of I against T is termed the glow curve and, in the idealised case of a single trapping level, may take the shape of any of the computed curves shown in figure 2.4. Initially, at low values of T the curves rises exponentially and after reaching a maximum—the glow peak—it falls to zero. The greater the value of E and the smaller the value of s, the higher is the temperature of the glow peak maximum, and hence the greater is the thermal stability of the trapped electrons. Either the peak height or the total area under the glow curve may be taken as a measure of the original ionising radiation exposure. The glow peak moves to lower temperatures as the value of R, the heating rate, is increased.

The theoretical mechanism discussed so far has related only to elec-

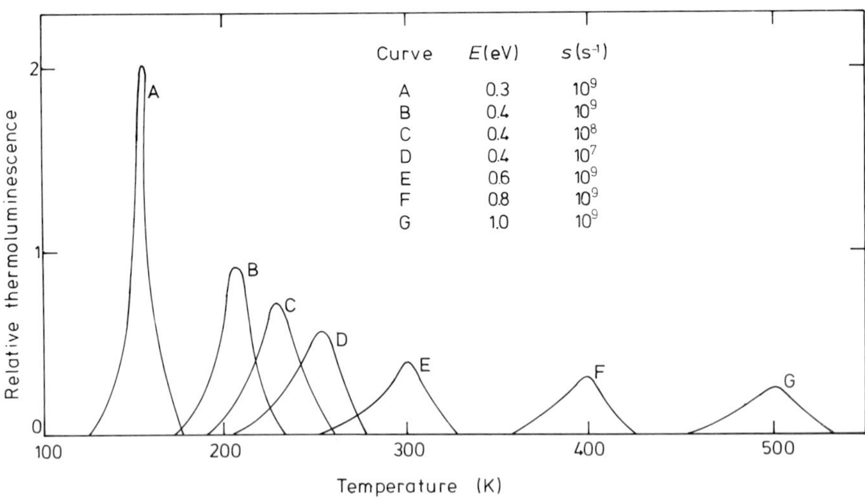

Figure 2.4 Theoretical glow curves for phosphors with single trap depth E and frequency factor s (Garlick *et al* 1949, reprinted with the permission of the Clarendon Press, Oxford.)

trons trapped at a single trapping level. In real phosphors many different trapping levels will be present, each one due to a particular lattice defect or complex of defects. Each trapping level will give rise to an associated glow peak maximum which may or may not be resolved during readout. The area and peak height of each glow peak depends on the number of associated electron traps present. This in turn depends on the number of lattice defects and, for real phosphors, on the type and amount of impurity atoms present, as well as on the thermal history and treatment of the material.

2.5 The Thermoluminescence Process in Lithium Fluoride

2.5.1 Introduction

The thermoluminescence phosphor most widely and intensively studied is lithium fluoride doped with magnesium and titanium—LiF : Mg : Ti. Many difficulties have arisen in understanding the theoretical significance of results of experiments performed on this material, primarily because of uncertainties in standardisation of phosphor composition. However, it is clear that the thermoluminescence process in LiF : Mg : Ti is complex and is critically dependent on a number of factors including: the amount and type of impurities present; the amount and type of intentional dopants present, their chemical form and method of introduction into the lattice; and the thermal, optical and mechanical treatment of the phosphor during its manufacture and use.

2.5.2 The role of magnesium ions

If LiF : Mg : Ti is given a pre-irradiation anneal at 400 °C (standard phosphor anneal of one hour) and cooled quickly to normal ambient temperature, the resulting glow curve after irradiation contains at least six glow peaks between normal ambient temperatures and 300 °C, as illustrated in figure 2.5. By convention these are named peaks 1 (60 °C), 2 (120 °C), 3 (170 °C), 4 (190 °C), 5 (210 °C) and 6 (285 °C). The precise readout temperature and the resolution of each peak will depend on the heating rate employed. Peak 5 is the one normally used for practical dosimetry. The low-temperature peaks exhibit high fading of stored signal even at normal ambient temperature; the half-life of each peak at normal ambient temperature is also shown in figure 2.5. It is possible, however, effectively to reduce the number of electron traps with which the low-temperature peaks are associated by thermally annealing the material for 1–2 h at 100 °C or 16–24 h at 80 °C prior to irradiation. This procedure results in the much more satisfactory glow curve also shown in figure 2.5.

The magnesium ions are presumed to form electron traps in combination with certain defect centres in the lattice. The influence of titanium in the trapping process is unclear and its role is thought to be primarily in the formation of luminescence recombination centres.

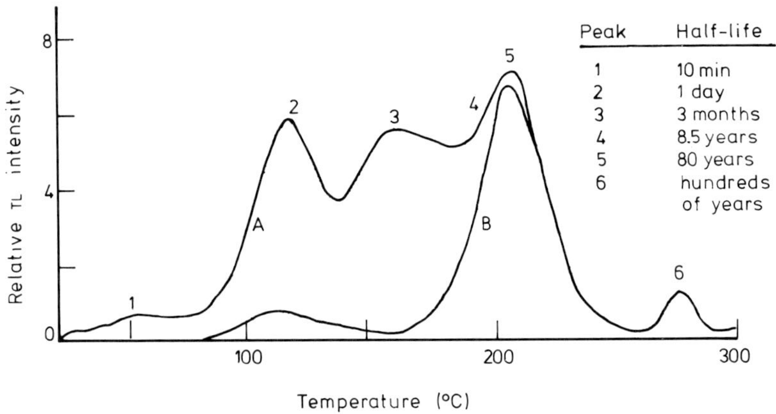

Figure 2.5 Glow curves for LiF:Mg:Ti (TLD 100) annealed for 1 h at 400 °C followed by: A, cooling (10^3 °C min^{-1}) to normal ambient temperature; B, 16 h anneal at 80 °C, followed by irradiation. The approximate value of the half-life of each peak is also shown (Mason *et al* 1976).

The divalent magnesium ion (Mg^{2+}) is introduced into a lattice consisting of an array of monovalent lithium (Li^+) and fluorine (F^-) ions. The substitution of Li^+ ion by a Mg^{2+} ion results in an excess positive charge at the lattice site. Coulombic attraction results in the formation of nearest-neighbour pairs (dipoles) consisting of a substitutional Mg^{2+} ion in combination with a Li^+ ion vacancy, as illustrated in figure 2.1(*b*). Under certain thermal conditions the dipoles are believed to aggregate forming dimers, trimers and higher-order complexes. There is evidence that the simple dipole arrangement is associated with the electron trapping centres responsible for the low-temperature glow peaks 2 and 3 (figure 2.5). Dipole complexes are associated with the main dosimetry glow peaks 4 and 5. The aggregation of simple dipoles to form complexes is critically dependent on temperature, and heating and cooling rates. Dipoles are electrically neutral in the lattice and the actual electron trapping centres may be akin to modified F centres such as a dipole in proximity to a fluorine ion vacancy (Mayhugh *et al* 1968). This model for trapping can be used to explain the changes in the relative heights of the glow peaks due to pre-irradiation annealing of the phosphor.

The standard pre-irradiation thermal annealing of TLD 100 phosphor at 400 °C followed by cooling to normal room temperature produces a glow curve with six peaks within the temperature range normal ambient temperature to 300 °C, as shown in figure 2.5. Indeed this is true for any anneal temperature above about 180 °C. With increasing anneal temperature, simple complexes such as dimers and trimers are broken up into dipoles, producing the relatively large peaks 2 and 3. If the phosphor is then annealed at a temperature $\lesssim$100 °C, such as for 1 or 2 h at 100 °C or 16–24 h at 80 °C, aggregation occurs enhancing the main dosimetry peaks 4 and 5 at the expense of the low-temperature peaks. Further evidence for aggregation is provided by examining the effect on the glow curve of the cooling rate from 400 °C to normal room temperature (Mason *et al* 1976). If the phosphor is cooled rapidly a number of associated trapping centres are 'frozen' into the lattice resulting in a relative enhancement of peak 2, as shown in figure 2.6. A slower cooling rate allows aggregation which relatively enhances peaks 4 and 5. Very slow cooling rates result in the formation of higher-order complexes and Mg^{2+} precipitation (Bradbury and Lilley 1977) and a reduction of peaks 4 and 5. It has been suggested that peaks 1 and 6 are not directly connected with the presence of either magnesium or titanium but possibly with intrinsic lattice defects (Crittenden *et al* 1974). However, Attix (1975) has suggested that peak 6 may be associated with magnesium dipole trapping centres which have trapped more than one electron.

2.5.3 The role of titanium and hydroxyl ions

The presence of titanium in combination with hydroxyl (OH) ions is a prerequisite for efficient thermoluminescence emission in LiF (Rossiter *et al* 1971). All crystalline materials grown under air from the melt contain OH ions in concentrations of several tens of parts per million. The thermoluminescence sensitivity of LiF:Mg:Ti has been shown to increase with increasing concentration of titanium ions present to a maximum content of 7 ppm (Rossiter *et al* 1971). Where high concentrations of titanium exist the thermoluminescence sensitivity appears to be controlled by low OH ion concentration and vice versa (Vora *et al*

1975, Rossiter *et al* 1971). The results of ionic conductivity experiments indicate that titanium may be present in the divalent state Ti^{2+}, possibly forming Ti^{2+}–Li^{+} vacancy dipoles in a similar manner to Mg^{2+} (Bloch 1968). As the concentration of titanium appears to change the relative shapes of the glow peaks it may also be involved in an unknown way in the formation of electron trapping centres (Niewiadomski 1976).

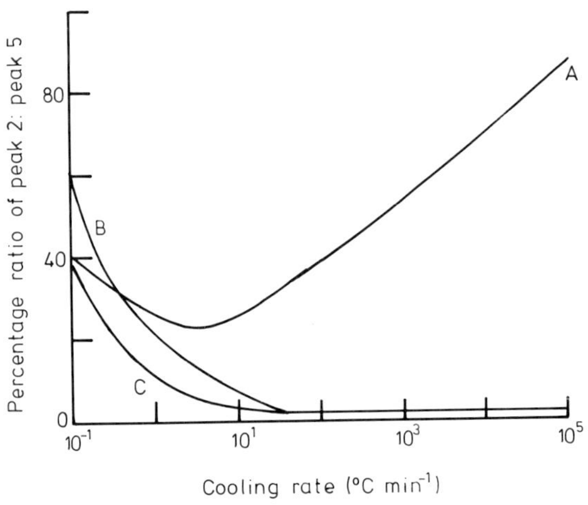

Figure 2.6 LiF:Mg:Ti. Percentage ratio of peaks 2:5 as a function of cooling rate from 400°C to normal ambient temperature for different anneal regimes. A, no anneal; B, 100°C anneal; C, 80°C anneal (Mason *et al* 1976).

2.5.4 Models for TL in LiF:Mg:Ti

Many models have been suggested for the thermoluminescence process in LiF:Mg:Ti and in recent years two models have proved particularly useful in explaining some of the experimentally observed phenomena. These are the models of Mayhugh (1970) and Nink and Kos (1976).

2.5.4.1 The Mayhugh model. On the basis of theory and experimental work on the observed changes in the optical absorption spectra of LiF crystals containing different amounts of magnesium, subjected to various irradiation, thermal and optical bleaching procedures, together with glow peak behaviour, Mayhugh postulated a model for trapping and recombination processes in LiF (Mayhugh 1970). The electron and hole energy transitions and relevant trapping levels are illustrated in figure 2.7.

In LiF:Mg:Ti at normal readout temperatures (> 200°C) electrons are released from the traps associated with glow peaks 4 and 5 (step 1). The electrons migrate through the lattice, in energy terms via the conduction band (step 2), and are finally captured at the multiple hole

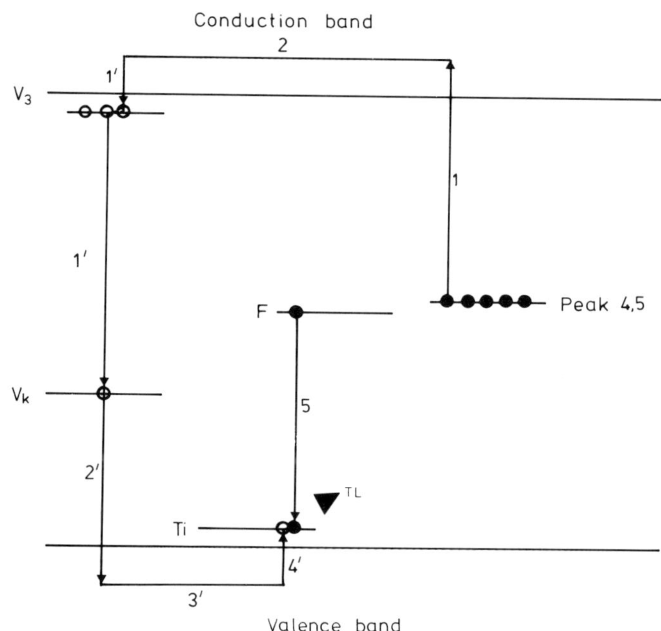

Figure 2.7 Mayhugh's model for the thermoluminescence process in LiF:Mg:Ti (Mayhugh 1970).

trapping centre V_k (step 1′). This centre is thermally unstable around normal ambient temperature and holes are immediately released into the lattice (step 2′). The holes migrate through the lattice (step 3′) and ultimately undergo thermoluminescence emissive recombinations at a titanium-associated 'activator centre' (step 4′) with an electron which has tunnelled from an F centre (step 5).

2.5.4.2 The Nink and Kos *Z* centre trapping model. By examining the comparative effects of different post-irradiation thermal annealing treatments on the optical absorption spectra of LiF:Mg (300 ppm) and

LiF:Mg (300 ppm): Ti (3 ppm), Nink and Kos (1976) have developed a model for electron trapping in LiF:Mg:Ti based on the irradiation and thermal conversion of Z_n centres. The structure of Z_n centres is shown in table 2.1.

Table 2.1 Structure of centres relevant to the Z_n centre model for the TL process in LiF.

Centre	Structure	Description
α	⊟	Negative ion vacancy
F	⊟	Negative ion vacancy plus one trapped electron
F′	⊟	Negative ion vacancy plus two trapped electrons
Z_0	Mg^{2+}–α	Magnesium ion plus α centre
Z_2	Mg^{2+}–F′	Magnesium ion plus F′ centre
Z_3	Mg^{2+}–F	Magnesium ion plus F centre

LiF:Mg:Ti annealed at 400 °C contains only Z_0 magnesium-related defects. These act as building blocks for the production of Z_2 and Z_3 centres using a supply of electrons (e) produced by ionising radiation thus:

$$Z_0 + e \rightarrow Z_3 \qquad (2.9)$$

$$Z_3 + e \rightarrow Z_2. \qquad (2.10)$$

At normal ambient temperature the irradiated crystal now contains Z_0, Z_2 and Z_3 centres.

During readout some of the Z_2 centres thermally decay into Z_3 centres releasing electrons thus:

$$Z_2 \rightarrow Z_3 + e, \qquad (2.11)$$

while the more thermally stable Z_3 centres remain unchanged. A proportion of the released electrons undergo luminescence recombinations.

2.6 Supralinearity and Sensitisation

2.6.1 Introduction

The TL per unit absorbed dose response is an important characteristic of a thermoluminescence phosphor; figure 2.8 illustrates a typical response.

At relatively low values of absorbed dose the response is linear

$$\mathrm{TL} = fD + B, \tag{2.12}$$

where TL is the thermoluminescence signal (peak height or integrated signal) for an absorbed dose D, f is the TL emitted per unit absorbed dose, and B is the background TL signal from an unirradiated phosphor.

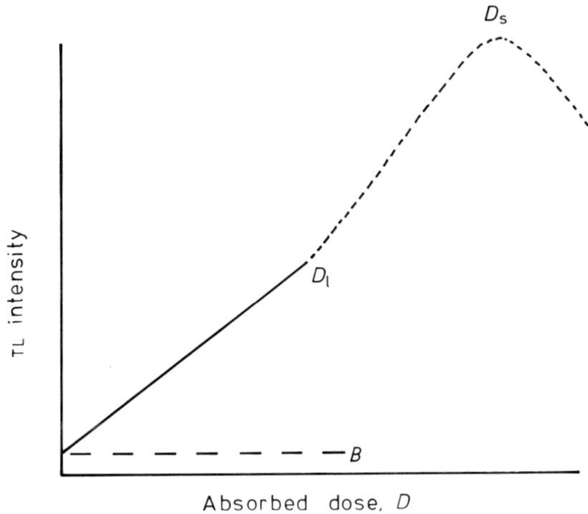

Figure 2.8 Typical TL–absorbed dose response curve for a TL phosphor.

The detection threshold of a dosemeter is a function of the statistical variability of the background signals obtained from a reasonably sized sample (say 50) of similar unirradiated dosemeters. It is usually defined as two or three times the standard deviation of a set of such backgrounds. In the linear region

$$\frac{\Delta(\mathrm{TL})}{\Delta D} = f. \tag{2.13}$$

Above D_l the response is supralinear, saturating at D_s and then falling off rapidly.

An absorbed dose of approximately 20% less than D_s is generally taken to be the practical upper absorbed dose limit. The supralinear response of a phosphor might appear to be a bonus, but increased sensitivity is not generally required at high absorbed dose levels. As further calibration has to be done, and correction factors measured and applied, the opportunity for error is increased. While the TL–absorbed dose response curve shown in figure 2.8 is generally appropriate to TL phosphors, some may display marked nonlinearity even at low levels of absorbed dose. For some phosphors the value of absorbed dose at which the onset of supralinearity occurs is dependent on factors such as the impurity ion concentration of the phosphor (for example, OH^- ions) and the linear energy transfer (LET) of the radiation.

It is found that if LiF : Mg : Ti phosphor is irradiated to a high absorbed dose (supralinear to saturation) and then annealed at around 300 °C for a short period, the thermoluminescence sensitivity of the material increases by a factor of up to five (Wilson *et al* 1966). This process is called sensitisation. A 1 or 2 h 400 °C anneal effectively reduces the sensitivity.

2.6.2 Models for supralinearity

Several models have been proposed to explain supralinearity and have usually been based on the following.

(1) An increase in the number or competitiveness of electron traps (Cameron *et al* 1965, Suntharalingam and Cameron 1969).

(2) An increase in the number or competitiveness of luminescence centres (Claffy *et al* 1968, Mayhugh *et al* 1968, Mayhugh 1970, Stoebe and Watanabe 1975, Attix 1975).

(3) A combination of (1) and (2).

Two models will be considered.

2.6.2.1 The track interaction model (Claffy *et al* 1968, Attix 1974, 1975). This general model can in principle be applied to any TL phosphor since it is not concerned with the specific nature of the electron traps or luminescence centres, but rather with their spatial distribution in the phosphor (figure 2.9). When a phosphor is irradiated the electron–hole pairs produced will be initially distributed along the direction of the secondary charged particles. Also, luminescence centres are assumed to be produced along these tracks as shown. At relatively low levels of absorbed dose only intra-track TL emissive recombinations are possible as the inter-track distances are large compared with the electron–hole migration distances. At relatively high levels of absorbed dose, the tracks are packed much closer together enabling inter-track recombinations. It is assumed that if the material is heated a significant number of the radiation-induced luminescence centres survive, thus sensitising the material. Attix also used this model to explain the observed loss of TL efficiency with increased linear energy transfer (LET) in lithium fluoride.

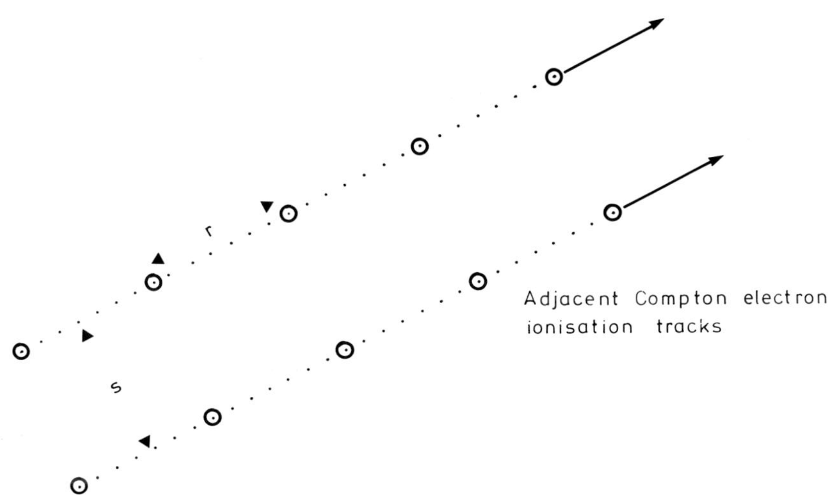

Figure 2.9 Track interaction model for supralinearity in a TL phosphor: · electron–hole pairs, ○ recombination (F) centres, r, recombination centre separation approx. 230 nm (intra-track electron–hole TL recombination distances approx. 60 nm); s, inter-track electron–hole TL recombination distance (approx. 200 nm at 1 Gy, i.e. linear response; approx. 66 nm at 10 Gy, i.e. onset of supralinearity; approx. 6.6 nm at 10^3 Gy, i.e. supralinear response) (Claffy *et al* 1968, Attix 1975).

2.6.2.2 The Mg^{2+}–OH^-–Li^+ vacancy model for supralinearity in LiF:Mg:Ti. The model proposed by Stoebe and Watanabe (1975) postulates the existence of Mg^{2+}–OH^-–Li^+ vacancy-associated centres which act as competitors with the normal luminescence emission centres, as illustrated in figure 2.10. During irradiation these traps capture electrons produced as a result of ionisation and will only partly fill at relatively low levels of absorbed dose. During normal readout, electrons released from peaks 1–5 associated traps may become retrapped or undergo the normal TL emission process at a luminescence centre. At relatively high levels of absorbed dose the number of electrons produced increases, as does the number released from peaks 1–5 associated traps,

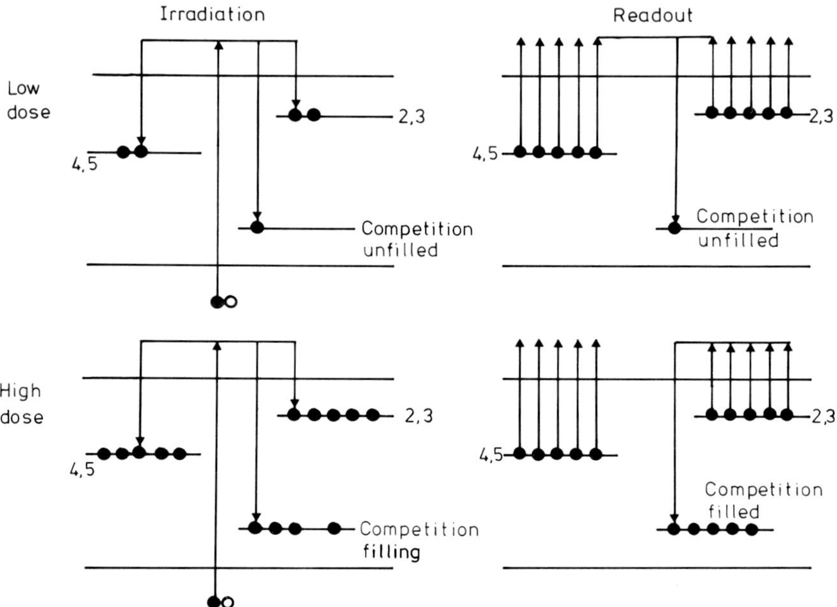

Figure 2.10 Supralinearity in LiF:Mg:Ti—competing trap model (Stoebe and Watanabe 1975).

filling up the competitive traps. There is an increase in the number of electrons available to undergo luminescence recombination resulting in an increased TL sensitivity at high levels of absorbed dose. In addition, during normal readout, the low-temperature traps release their trapped electrons first, thereby filling up the competitive traps first, resulting in the onset of supralinearity at a lower level of absorbed dose for the higher-temperature dosimetry traps and a larger sensitivity increase. This agrees with observation. It is assumed that a subsequent short 300 °C anneal only partly empties the competitive traps resulting in sensitisation, but a 400 °C anneal empties them completely returning the phosphor to its former state.

2.7 Phototransferred Thermoluminescence (PTTL) and Re-assessment of Absorbed Dose

The irradiation of a TL phosphor with ultraviolet radiation may stimulate the release of charge carriers (electrons or holes) from trapping centres

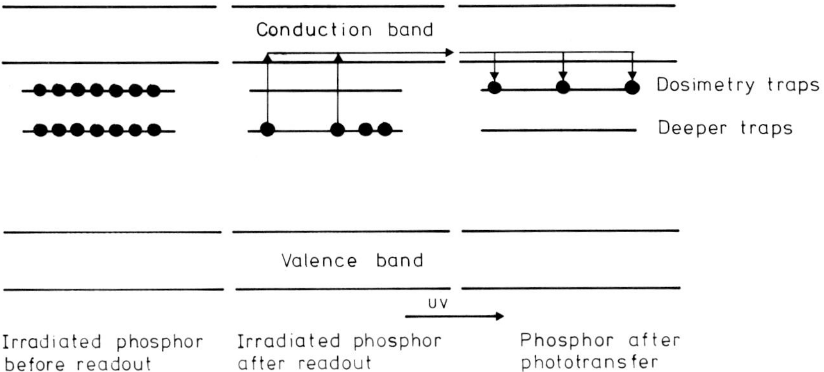

Figure 2.11 Energy band model for phototransferred thermoluminescence (PTTL).

where they are normally stable at normal maximum readout temperatures. The charge carriers released become available for retrapping by the main dosimetry traps, as illustrated in figure 2.11. In LiF:Mg:Ti the phototransfer retrapping of electrons is used to provide a re-assessment of absorbed dose in a phosphor which has already been read out

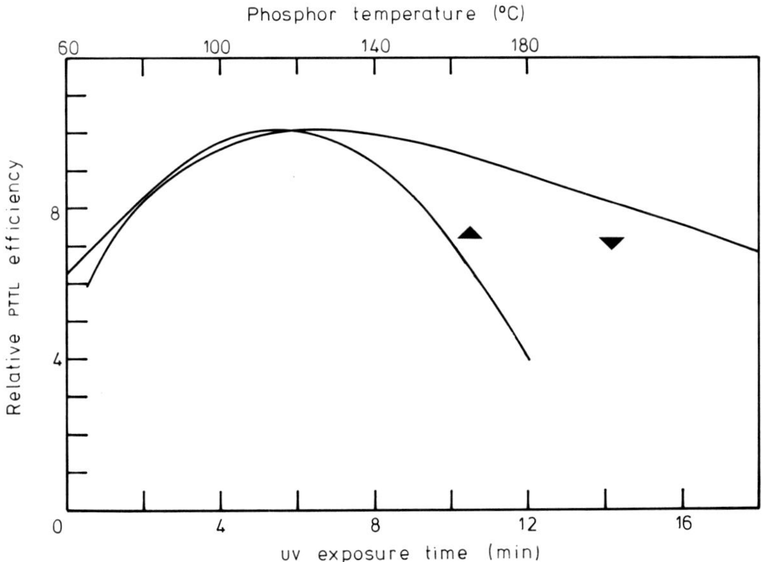

Figure 2.12 General dependence of PTTL efficiency on ultraviolet exposure and phosphor temperature for LiF:Mg:Ti. UV irradiance 4.0 mW cm^{-2}.

(Mason *et al* 1977). By repeating the normal readout procedure following exposure to ultraviolet radiation of 254 nm wavelength (mercury vapour resonance emission), the thermoluminescence (PTTL) measured is related to the original radiation absorbed dose. In LiF:Mg:Ti phosphor the optimum transfer efficiency of electrons from the deep traps to the dosimetry traps is a function of the total radiant exposure of ultraviolet radiation and the temperature at which it is carried out (Mason *et al* 1977, Bartlett *et al* 1980), as illustrated in figure 2.12.

3 The Characteristics of Thermoluminescence Materials

3.1 Introduction

In this chapter the important characteristics of thermoluminescence phosphors are examined, and those of different materials are reviewed and compared. The materials selected have been chosen on the basis of popularity of use, commercial availability and specific applications, including personal, clinical and environmental dosimetry. The most important families of phosphors included are lithium fluoride (LiF), lithium borate ($Li_2B_4O_7$), calcium sulphate ($CaSO_4$) and calcium fluoride (CaF_2). The properties of these and other less widely used phosphors are compared and tabulated. The characteristics of different physical forms of dosemeter are also examined.

3.1.1 Thermoluminescence–absorbed dose reponse

While not essential, it is extremely useful for a phosphor to have a linear TL–absorbed dose response over measurement and calibration ranges of absorbed dose. As discussed in § 2.6, the response of TL phosphors is usually, but not exclusively, linear at low absorbed dose values, becomes supralinear, and finally saturates at high values. Supralinearity leads to higher uncertainty in the measured absorbed dose since additional calibration factors have to be determined and applied. Dosemeters which have been used at high levels of absorbed dose and then re-used at low levels must be adequately annealed to remove any thermoluminescence memory. The intrinsic TL sensitivity of a phosphor may be expressed as the TL yield per unit mass of phosphor per unit exposure of ionising radiation, provided a correction is applied to take account of the spectral match of the TL emission and peak spectral sensitivity of the reader. The sensitivity of a phosphor may also change with form, e.g. different grain sizes, and often most markedly with photon energy or LET of the radiation. For all practical purposes the response of TL dosemeters appears to be independent of absorbed dose rate. For example, the response of LiF : Mg : Ti has been shown to be independent of absorbed dose rate up to 10^8 Gy s^{-1} (Tochilin and Goldstein 1966).

3.1.2 Relative photon energy response

The total TL emitted by an irradiated phosphor is proportional to the total radiation energy absorbed by it. The mass energy absorption coefficient of any thermoluminescence phosphor may be calculated from the mixtures rule:

$$(\mu_{en}/\rho)_{phosphor} = \sum_i (\mu_{en}/\rho)_i w_i,$$

where $(\mu_{en}/\rho)_i$ is the mass energy absorption coefficient of the ith elemental constituent of the phosphor, and w_i is the fraction of that element in the phosphor. The mass energy absorption coefficient of any element is a function of photon energy, and is dependent on the main photon absorption and other interaction processes, such as the photoelectric effect, Compton scatter, pair production and (of relatively minor importance) Rayleigh (elastic) scatter. Tables of theoretical mass energy absorption coefficients and photon interaction cross sections have been compiled by several authors (e.g. Storm and Israel 1967). All coefficients and cross sections are dependent on the atomic number Z of the target atoms, and on the photon energy E. As a TL phosphor comprises many atoms of the basic lattice, plus relatively few dopant atoms, the simultaneous absorption and scattering processes are complex. The total photon interaction cross section per atom, σ_{tot}, may be written as

$$\sigma_{tot} = \sigma_{pe} + \sigma_{cs} + \sigma_{pp},$$

where σ_{pe}, σ_{cs}, and σ_{pp} are the individual interaction cross sections for the photoelectric effect, Compton scatter and pair production, respectively. The approximate dependence of these interaction cross sections as functions of atomic number Z and photon energy E is shown in table 3.1.

For elements of low atomic number, and for photon energies up to approximately 15 keV, the photoelectric effect is dominant, but thereafter up to 10 MeV Compton scatter is important. In tissue the Compton process is the most important in the energy range approximately 20–10^4 keV. For elements of high atomic number, which TL dopant materials often are, the photoelectric process is dominant up to several hundred keV. The photon energy response of a TL phosphor may be expressed in different ways and a commonly used method is to compare the response of the phosphor normalised at a particular photon energy, often ^{60}Co gamma energy (1.25 MeV mean), with that of air or tissue. The relative energy response (RER) of the phosphor at photon energy E is

$$(\mathrm{RER})_E = \frac{[(\mu_{en}/\rho)_{\mathrm{phosphor}}/(\mu_{en}/\rho)_{\mathrm{air}}]_E}{[(\mu_{en}/\rho)_{\mathrm{phosphor}}/(\mu_{en}/\rho)_{\mathrm{air}}]_{1.25\ \mathrm{MeV}}}.$$

If $(\mathrm{RER})_E$ is plotted as a function of photon energy, plots of kerma per unit exposure are obtained normalised at 1.25 MeV. Under conditions of electronic equilibrium the plots are of absorbed dose per unit exposure.

Table 3.1 Approximate dependence of photon interaction cross sections on the atomic number (Z) of the absorber.

Process	Approximate dependence
Photoelectric effect	σ_{pe} varies as Z^4 for low-energy photons σ_{pe} varies as Z^5 for high-energy photons
Compton scatter	σ_{cs} varies as Z
Pair production	σ_{pp} varies as Z^2(>1.02 MeV)

3.1.3 Fading

After exposure of a TL phosphor to ionising radiation, the latent measure of the absorbed dose is the number of electrons which remain trapped in the various trapping levels. Unintentional release of these electrons before readout is called fading. Fading may be due to thermally or optically stimulated release of the electrons or a combination of both.

3.1.3.1 Thermal fading. As discussed in Chapter 2, the probability for the thermal release of an electron from a trapping centre is governed by Boltzmann's equation

$$p = s \exp(-E/kT).$$

The half-life ($T_{1/2}$) of a particular trap (and, by association, a particular glow peak) is defined as the time for the number of trapped electrons to fall to half of its original value

$$T_{1/2} = \frac{0.693}{p}\ \mathrm{s}.$$

Potentially high thermal fading is generally indicated by the presence of low-temperature glow peaks or a single, broad glow peak with a low-temperature component. A pre-irradiation anneal, such as used with LiF:Mg:Ti, or controlled cooling rates, will reduce the relative magnitudes of the low-temperature peaks. Also a post-irradiation anneal (pre-read), either in an oven or in the reader, can be used to fade intentionally any filled low-temperature traps before readout of the TL signal from the higher-temperature main dosimetry traps.

3.1.3.2 Optical fading. Exposure to photons of visible and particularly ultraviolet radiation can produce a twofold effect. In an unirradiated dosemeter this may cause a spurious TL signal. An example of this is LiF:PTFE dosemeters, in which spurious thermoluminescence equivalent to several mGy may be induced by prolonged exposure of the PTFE matrix to the ultraviolet component of natural or artificial light. In an irradiated phosphor it may additionally cause loss and/or redistribution of trapped dosimetry electrons by phototransfer, as discussed in § 2.7.

3.2 Lithium Fluoride

Introduction. Lithium fluoride (LiF) is an alkali halide with a density of 2.64 g cm^{-3} and photon effective atomic number $Z_{\mathrm{eff}} = 8.2$. It is reasonably resistant to chemical attack and is only slightly soluble in water.

LiF:Mg:Ti is currently the most commonly used family of thermoluminescence phosphors and was first investigated by Daniels *et al* (1953). The original material which was 'left over' from optical component manufacture had a relatively high TL sensitivity but it was found that subsequent samples of the phosphor had variable TL properties and so its use was abandoned in favour of alumina (Rieke and Daniels 1957). Cameron *et al* (1961) renewed interest in it and Harshaw Chemicals produced a commercial LiF phosphor known as TLD 100 and its isotopic variants TLD 600 and TLD 700. The relative proportions of ^{6}Li and ^{7}Li contained in these phosphors are listed in table 3.2 The use of the isotopic variants in the fields of neutron and mixed-field dosimetry is discussed in § 5.2.

TLD 100 (600, 700) is produced by homogeneous melting of lithium fluoride, magnesium fluoride, lithium cryolite and lithium titanium fluoride, resulting in a phosphor containing 300 ppm magnesium and 10–

20 ppm titanium. A single crystal is solidified from the melt, then pulverised, and the powder grains sieved and separated. Extruded-ribbon dosemeters (Cox 1968) are formed by compressing the original mixture at an elevated temperature, causing it to extrude through an aperture. The extrusion is cut into pieces which are then polished.

Table 3.2 Isotopic constituents of Harshaw TLD 100, 600 and 700.

Phosphor type	^{6}Li (%)	^{7}Li (%)
TLD 100	7.5	92.5
TLD 600	95.6	4.4
TLD 700	0.01	99.99

Portal *et al* (1971) have prepared a 'sodium-stabilised' lithium fluoride phosphor by the addition of 1–2% wt/wt sodium fluoride. The material can be re-used at normal low levels of absorbed dose with no thermal treatment other than an initial 500 °C anneal. The phosphor is available commercially as PTL 710, with isotopic variants PTL 716 and PTL 717.

The spectral emission peaks of both TLD and PTL phosphors are at 400 nm in the blue with full-width half-heights of approximately 115 nm, as shown in figure 3.1. This emission corresponds with the peak quantum efficiency of S11 and bi-alkali photomultipliers.

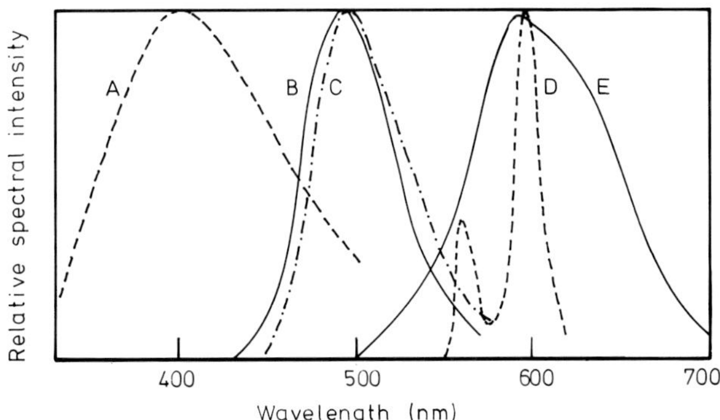

Figure 3.1 TL emission spectra of various phosphors. A, LiF:Mg:Ti (TLD 100); B, CaF_2:Mn; C, $CaSO_4$:Mn; D, $CaSO_4$:Sm; E, $Li_2B_4O_7$:Mn. (Curves A, B and C from Fowler and Attix 1966 *Solid State Integrating Dosemeters,* in *Radiation Dosimetry* vol II, 2nd edn, reprinted with the permission of Academic Press.)

Glow curve and TL–absorbed dose response. The glow curve of LiF : Mg : Ti phosphor has at least six peaks between normal ambient temperature and 300 °C, as illustrated in figure 2.5. By applying the standard annealing procedure described in § 2.5.2 the traps associated with the low-temperature peaks 1, 2 and 3 are reduced effectively in number. This necessary and somewhat complicated thermal anneal is regarded as a major disadvantage of the phosphor. The relative and absolute heights of the glow peaks depend on the LET of the radiation, and the ratio of the height of peak 6 (at 285 °C) to that of peak 5 (at 210 °C) has been used for mixed-field dosimetry, as discussed in § 5.2. The relative and absolute heights are also affected by the cooling rate from the high-temperature anneal, as illustrated in figure 2.6. For maximum sensitivity, TLD and PTL LiF materials should be cooled fairly rapidly and reproducibly.

Table 3.3 Comparison of relative TL sensitivities of a number of LiF-based phosphors (Driscoll 1977).

Phosphor type	Relative sensitivity
TLD 700†	1.00 ± 0.16
PTL 717‡	0.50 ± 0.07
BNFL 700§	0.77 ± 0.11
NPL 200‖	0.62 ± 0.06

† Harshaw Chemical Co. Inc., USA.
‡ Desmarquest and Carbonisation Enterprise et Céramique, France.
§ British Nuclear Fuels Ltd, UK.
‖ National Physical Laboratory, UK.

Driscoll (1977) compared some of the dosimetric characteristics of a number of LiF-based powdered phosphors including TLD 700, PTL 717, LiF 700 (British Nuclear Fuels), NPL 200 (National Physical Laboratory), and concluded that TLD 700 was the most sensitive, as shown in table 3.3. The lowest detectable absorbed dose depends on the physical form of the dosemeter and the readout equipment. For powdered and extruded forms of TLD 100 the threshold is usually between 50 and 100 μGy for ^{60}Co radiation using a normal commercial reader. The absorbed dose response is linear up to between 3 and 10 Gy depending on the form of the dosemeter. The onset and degree of supralinearity also depends on the LET of the radiation, as illustrated in

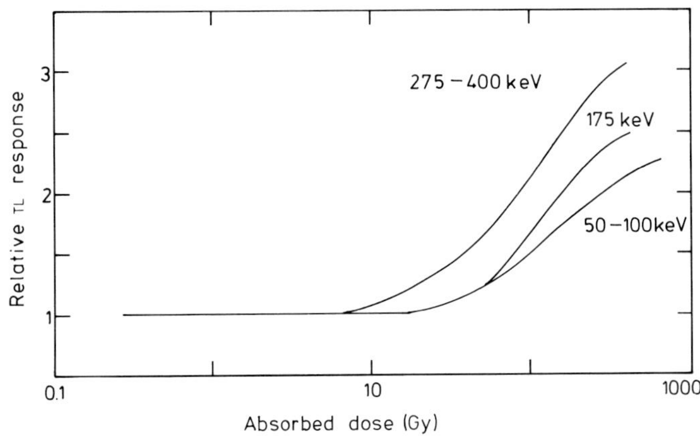

Figure 3.2 The relative response (TL per Gy) of LiF:Mg:Ti (TLD 100) extruded ribbons for electrons of various LET (Suntharalingam and Cameron 1969).

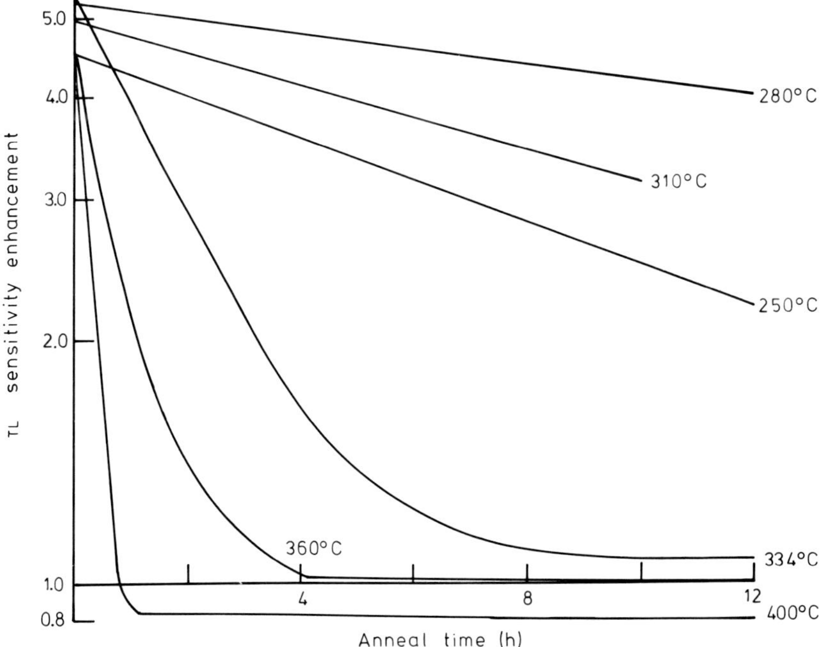

Figure 3.3 Sensitivity enhancement in TLD 100 following an exposure of 3×10^4 R of ^{137}Cs gamma rays and various post-irradiation anneals (Wilson *et al* 1966).

figure 3.2, and on the presence of OH^- ions. A sensitising absorbed dose of between 300 and 10^3 Gy followed by a 1–4 h anneal at 280 °C results in a fivefold sensitivity enhancement and a reduction in supralinearity. If the phosphor is subsequently annealed at temperatures > 360 °C for longer than a few hours the enhancement is lost; indeed for anneals of longer than 2 h at temperatures $\gtrsim 400$ °C the sensitivity falls below its original value (figure 3.3).

Very high absorbed doses ($> 10^4$ Gy) produce permanent radiation damage in the lattice and an irreversible loss of sensitivity.

Relative photon energy response. LiF has a photon effective atomic number $Z_{\text{eff}} = 8.2$, compared with 7.4 for tissue, and for most applications it can be considered to be approximately tissue-equivalent with a maximum over-read of 1.3 to 1.4 at 20–30 keV. The relative photon energy response of LiF is illustrated in figure 3.4.

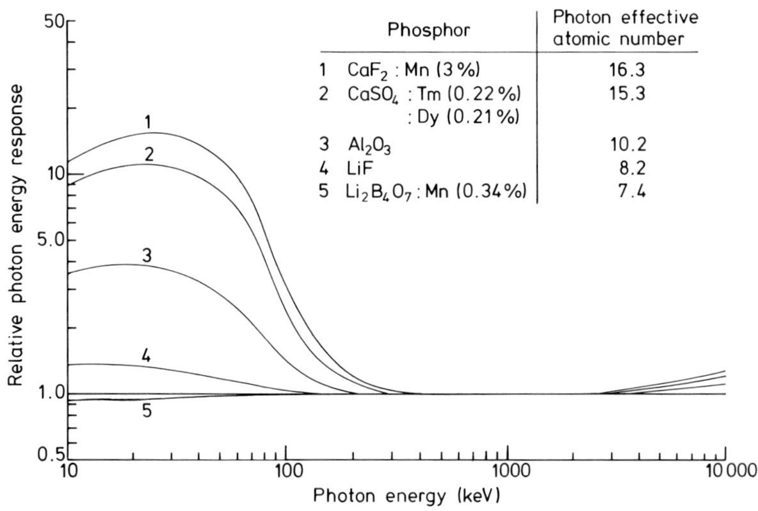

Figure 3.4 Theoretical photon energy responses of a number of TL phosphors relative to that of air (1.00). After Bassi *et al* (1976).

Fading. The fading of the low-temperature peaks of the glow curve of LiF:Mg:Ti is high; half-lives of the various peaks at ambient temperature are listed in figure 2.5.

Because of the presence of so many different traps the fading process is undoubtedly complex. Booth *et al* (1972) observed the apparent transfer of TL sensitivity from low-temperature traps (peaks 2 and 3) to the main dosimetry traps (peaks 4 and 5) after the high-temperature anneal, prior to irradiation. Such an effect depends on the relative proportions of the traps present, which are determined by thermal treatment, especially cooling rates, as shown in figure 2.6. Mason *et al* (1976) could not explain the results of Booth *et al* for normal cooling rates ($< 10^3$°C min^{-1}) encountered in routine oven and reader anneals. With the application of the appropriate pre-irradiation anneal and/or post-irradiation pre-read, the apparent fading of the stored signal and TL 'sensitivity transfer' may be reduced to a negligible minimum for all practical absorbed dose measurement purposes. The relative fading of lithium fluoride TLD 700 extruded-ribbon dosemeters for a variety of post-irradiation storage temperatures between 5 and 100°C is shown in figure 3.5.

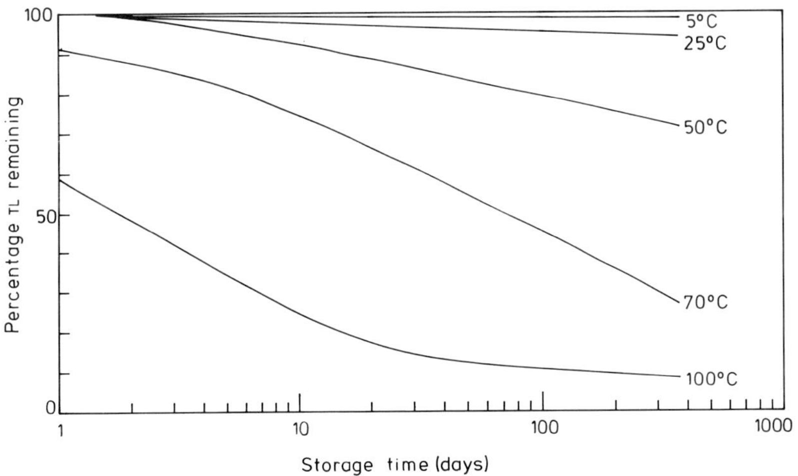

Figure 3.5 Thermal fading of LiF:Mg:Ti chips for storage temperatures between 5 and 100°C. 100°C, 20 min pre-read anneal (Burgkhardt *et al* 1977).

3.3 Lithium Borate

Introduction. Lithium borate ($Li_2B_4O_7$) phosphor doped with 0.1% wt/wt manganese, was introduced by Schulman *et al* (1965). It had the advantages of a low effective atomic number, $Z_{eff} = 7.4$ (slightly lower than that of air), and hence good photon energy tissue equivalence; a simple glow peak structure with a composite low-temperature peak between 55 and 90 °C and a dosimetry peak at 200 °C; a relatively simple preparation technique; low production costs; and a straightforward thermal regeneration procedure (15 min at 300 °C for heavily irradiated samples). Its major disadvantage was that despite its estimated high intrinsic TL efficiency (0.073 compared with 0.04 for LiF), its overall sensitivity was only about one tenth of that of lithium fluoride due to its yellow/orange spectral emission at 600 nm (Gorbics 1965), in contrast with the more efficiently detectable 400 nm blue emission of LiF (see figure 3.1).

The TL–absorbed dose response was linear from approximately 0.001 to 1.5 Gy with saturation at 3000 Gy. Other investigators have examined the effects on TL sensitivity, glow curve shape, fading, photon energy response, physical stability, etc, of varying the preparation parameters, including the concentration of manganese, the relative proportions of Li_2O and B_2O_3, and the duration and temperature of melt, etc (Kirk *et al* 1967, Christensen 1967, 1968, Wallace and Ziemer 1968, Jayachandran 1970, Brunskill 1968, Böttor-Jensen and Christensen 1972). Physical deterioration of the dosemeter and loss of TL sensitivity associated with the combined effects of temperature and humidity have been observed (Mason *et al* 1974, Burgkhardt *et al* 1977).

The spectral emission of manganese (600 nm) remained a problem, however, and attempts were made to introduce a dopant which would emit in the blue or ultraviolet. Thompson and Ziemer (1973) substituted silver for manganese and produced a phosphor with a peak emission at 290 nm in the ultraviolet but, unfortunately, the overall sensitivity of the phosphor was unchanged due to the apparent low efficiency of the silver activator. Attempts to use rare earth dopants have not as yet produced a sensitive phosphor (e.g. Rzyski and Nambi 1977), although Takenaga *et al* (1977), using Schulman's method of preparation, have recently produced a most promising phosphor with co-activators copper and silver (0.02% wt/wt). This phosphor has two glow peaks, a low-temperature peak at 110 °C and a dosimetry peak at 185 °C. The emission spectrum has peaks at 268 and 530 nm associated with the silver and at 368 nm associated with the copper. Its TL sensitivity is high and the TL–absorbed dose response is linear from 30 μGy to 5 Gy. However, although the 'dark' thermal fading at normal ambient temperature is low ($<$ 9% after 60 days), the optically induced fading is very high—80% after 7 h exposure to normal ambient fluorescent lighting conditions.

Glow peaks and TL–absorbed dose response. The temperature and relative magnitudes of the glow peaks of lithium borate phosphors depend on the type and amount of activator, the phases of the Li_2O–B_2O_3 system present, and the thermal and other treatment of the phosphor. A summary of the glow peak structure and TL–absorbed dose response data for a few selected phosphors is presented in table 3.4.

Table 3.4 Summary of glow peaks and absorbed dose responses of lithium borate phosphors.

Activator	Glow peak temperature (°C)	Linear absorbed dose range (Gy)
0.1% Mn	55–100, 195–240†	10^{-4}–3†
0.1% Mn (glass)	95	0.3–10^3
0.1% Ag	100, 175	10^{-2}–3‡
0.02% Ag 0.02% Cu	110, 185	3×10^{-5}–5

† Temperatures of glow peaks and TL sensitivity depend on the preparation technique. The highest sensitivity was obtained using a red-response PM tube.

‡ Minimum experimental absorbed dose investigated.

Relative photon energy response. The effective atomic number (Z_{eff}) of lithium borate is 7.4. Christensen (1967) suggested 0.45% manganese for a phosphor with an air-equivalent response, the maximum variation being ±5% in the photon energy range 10–1000 keV. Jayachandran (1970) showed that by altering the amount of manganese the photon energy response could be made to match that of soft tissue (0% Mn), water (0.1% Mn), ICRU striated muscle (0.25% Mn), and air (0.34% Mn). The relative photon energy response of lithium borate phosphor is shown in figure 3.4.

Fading. The low-temperature peak(s) of lithium borate phosphors fade rapidly at normal ambient temperature, as shown in figure 3.6, but the effect on TL stability can be eliminated by a suitable post-irradiation anneal, such as for 3 min at 80 °C, or by a pre-read in the reader. The dosimetry peak of Mn-doped lithium borate is reasonably stable at normal ambient temperature; Christensen (1968) measured fading of 10, 25 and 37% after storage for 2, 6 and 13 months, respectively (figure 3.6). Lithium borate is significantly adversely affected by humidity.

Storage at relative humidities > 30% results in a significant loss of TL sensitivity in both irradiated and unirradiated phosphors indicating the possible destruction of luminescence centres as a result of thermal diffusion into the lattice of hydroxyl ions, oxygen ions or water molecules. The loss of sensitivity increases with increasing relative humidity and temperature (Mason *et al* 1974). The effect is not limited to powder but has been shown to occur also in lithium borate : PTFE discs. The 'dark' thermal fading at normal ambient temperatures of Ag : Cu-doped lithium borate is similar to that of the Mn-doped phosphor (< 9% in 2 months), but the effect of exposure to normal ambient levels of fluorescent lighting is dramatic (figure 3.6), necessitating light, tight packaging.

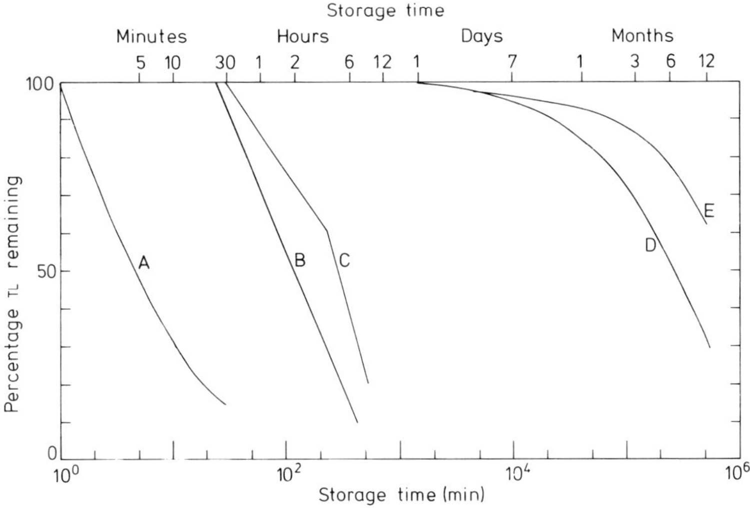

Figure 3.6 Thermal fading of selected $Li_2B_4O_7$ phosphors at normal ambient temperatures. A, $Li_2B_4O_7$:Mn low-temperature peak (Brunskill 1968); B and C, $Li_2B_4O_7$:Ag:Cu 1500 and 600 lux illumination respectively (Takenaga *et al* 1977); D, $Li_2B_4O_7$:Mn:PTFE 50% relative humidity (Burgkhardt *et al* 1977); E, $Li_2B_4O_7$:Mn (Christensen 1968).

3.4 Calcium Sulphate

Introduction. The calcium sulphate ($CaSO_4$) family of phosphors, together with calcium fluoride (CaF_2), was one of the earliest thermoluminescent materials studied and over the past 80 years it has often been used to measure the intensity of ultraviolet radiation. More recently, Watanabe (1951) was the first to prepare synthetic calcium sulphate doped with manganese for ionising radiation dosimetry. Since then several phosphor preparations using a variety of dopants have been studied, and four principal phosphors have emerged: those doped with manganese, samarium, dysprosium and thulium.

Calcium sulphate has a high effective atomic number, $Z_{eff} = 15.3$, and therefore its relative photon energy response is highly non-uniform. Its TL sensitivity per unit exposure at 30 keV is approximately 11 times that at ^{60}Co energy (figure 3.4). However, in its several forms it is one of the most sensitive phosphors available.

Although $CaSO_4$:Mn is the most sensitive of all the forms (Lippert and Mejdahl 1965 measured 5 μR with a standard deviation of 10%), it has not found universal application principally because of its very high thermal fading. It has a single glow peak at between 90 and 160 °C, depending on the heating rate, and a spectral emission peak at 500 nm (figure 3.1). It is light-sensitive, exposure to daylight is equivalent to a gamma photon exposure rate of about 15 μR min^{-1} with an equilibrium level of a few mR (Lippert and Mejdahl 1965).

In an attempt to reduce the thermal fading, samarium (Sm) was substituted for manganese (Kraysnaya *et al* 1961, Bjarngard 1965) producing a phosphor with a very stable glow peak at 400°C and a low-temperature glow peak at 100°C, but with an unfavourable spectral emission at 600 nm (figure 3.1). This combination of high-temperature glow peak and red emission limits the practical threshold of absorbed dose measurement to 10 mGy. By far the most popular $CaSO_4$ phosphors at present are those doped with dysprosium (Dy) and thulium (Tm) first produced by Yamashita *et al* (1971). Their main dosimetry glow peak is at 220°C, with two unstable low-temperature peaks at 80 and 120°C. At absorbed dose levels above the limit of linearity (i.e. 3 Gy for Tm- and 30 Gy for Dy-doped phosphors) a higher-temperature peak appears at approximately 250°C. This peak dominates up to the saturation absorbed dose of 100 Gy for Tm and 10^3 Gy for Dy. The spectral emission peak of the Tm-activated phosphor is at 452 nm, and at 478 and 571 nm for Dy. It has been shown that Dy gives essentially the same emission spectrum regardless of the host lattice.

As with Sm-doped material, both Tm- and Dy-activated phosphors produce spurious TL when exposed to sunlight. Yamashita *et al* (1971) measured the TL equivalent signal of 100 μGy after exposure of the phosphor to summer sunlight for 1 h. However, the effect of tribothermoluminescence is lower than for LiF : Mg : Ti phosphor. $CaSO_4$: Mn, $CaSO_4$: Tm and $CaSO_4$: Dy phosphors are all commercially available.

Glow curve and TL–absorbed dose response. A summary of the glow peak structure and TL–absorbed dose response characteristics of variously doped calcium sulphate phosphors are presented in table 3.5.

Table 3.5 Summary of glow peaks and absorbed dose response of calcium sulphate phosphors.

Activator	Glow peak temperature (°C)	Linear absorbed dose range (Gy)
Mn	90–160†	5×10^{-8}–10‡ saturates at 500
Sm	100, 400	10^{-2}–10^{2}§
Dy	80, 120, 220, 250‖	$<2 \times 10^{-6}$–30 saturates at 1000
Tm	80, 120, 220, 250‖	$<2 \times 10^{-6}$–3 saturates at 100

† Depends on heating rate.
‡ Nonlinear at absorbed doses <10 μGy.
§ Highly nonlinear response.
‖ Appears only at high absorbed doses.

Relative photon energy response. The effective atomic number of calcium sulphate phosphor is $Z_{eff} = 15.3$, and therefore the phosphor is not tissue-equivalent. However, its photon energy response almost matches that of bone whose effective atomic number is 14.3. Its relative energy response is shown in figure 3.4.

Fading. With its single low-temperature glow peak at 90 °C, $CaSO_4$: Mn exhibits a very high thermal fading at normal ambient temperature : about 50–60% after 24 h, making it unsuitable for many dosimetry applications (figure 3.7).

The fading of the high-temperature peaks of $CaSO_4$: Sm is negligible at storage temperatures up to 200 °C, enabling absorbed dose measurements at high ambient temperatures. At normal ambient temperature the 'dark' fading of the 220 °C glow peaks of the Tm- and Dy-activated phosphors is low, as shown in figure 3.7. However, exposure to direct sunlight for a few hours causes fading of between 3 and 30%, depending on the intensity of the light.

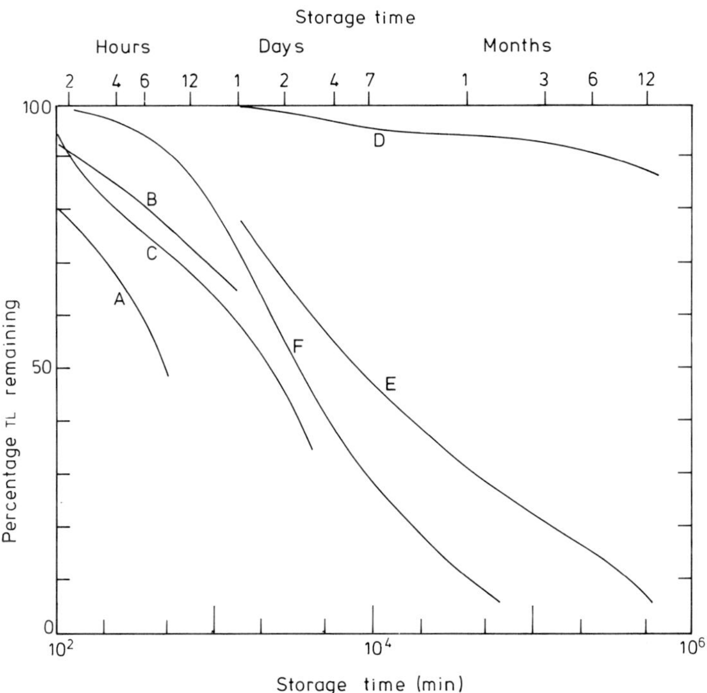

Figure 3.7 Thermal fading of $CaSO_4$ phosphors. A and B, $CaSO_4$:Mn 37 and 25 °C respectively (Bjarngard 1965); C, $CaSO_4$:Mn 26 °C (Lippert and Mejdahl 1965); D and E, $CaSO_4$:Dy:PTFE (0.4 mm) 25 and 100 °C respectively (Burgkhardt *et al* 1977); F, $CaSO_4$:Sm calculated relative fading of 400 °C glow peak for phosphor exposed at low exposure rate at 250 °C (Bjarngard 1965). (D and E, pre-read anneal of 20 min at 100 °C.)

3.5 Calcium Fluoride

Introduction. Like calcium sulphate, calcium fluoride (CaF_2) is not tissue-equivalent, and has a photon effective atomic number $Z_{eff} = 16.3$. The over-response at 30 keV compared with ^{60}Co energy is approximately 15 times. However, in all its forms it is an extremely sensitive dosimetry material and generally displays good TL–absorbed dose response linearity. The natural mineral form of calcium fluoride is fluorite, and its thermoluminescence properties were studied as early as the beginning of this century. The application of fluorites to ionising radiation dosimetry was reported by Schayes *et al* (1965). They have many advantages: they are cheap, abundant in nature, and, although different fluorites exhibit a complex array of glow peaks, most have thermally stable dosimetry peaks at approximately 260 and 275 °C. Lower-temperature, thermally unstable peaks are also present however. Their TL emission is in the ultraviolet (UV), blue and green regions of the spectrum.

They can be read out or annealed at temperatures up to 450 °C without affecting their sensitivity, and a gamma-ray absorbed dose of approximately 1000 Gy, followed by a 30 min anneal at 450 °C, results in an increase in TL sensitivity of more than 60%. The materials are, however, very sensitive to light and particularly also to ultraviolet, producing a twofold effect. A spurious TL signal is induced in the dosimetry peaks and electrons may be phototransferred to the dosimetry traps from deep traps associated with glow peaks at readout temperatures greater than 500 °C. The latter effect enables re-assessment of absorbed dose.

Ginther and Kirk (1957) and Schulman *et al* (1960) developed Mn-activated CaF_2 phosphor. The TL sensitivity of this material is high, although not as high as that of the fluorites. In contrast with the natural fluorites its glow peak structure is relatively simple. The dosimetry peak is a single composite peak at 260 °C. There are in addition higher-temperature peaks. The 'dark' fading at normal ambient temperature is surprisingly high (figure 3.8), and the TL emission peak is in the green region at 500 nm (see figure 3.1).

Binder *et al* (1968) introduced Dy as a dopant material. This phosphor has a somewhat complex glow curve structure with thermally unstable peaks at approximately 120 and 140 °C, peaks at 200 and 240 °C, and high-temperature peaks at 340 and 400 °C. As a result of the differential growth of each glow peak with increasing total absorbed dose, the TL–absorbed dose response is also complex. However, after annealing the phosphor for 2 h at 600 °C, or 3 h at 500 °C, the 140 °C peak has a linear response up to approximately 10^3 Gy with saturation at approximately 10^4 Gy. This thermal treatment reduces the overall sensitivity of the material by a factor of about two. CaF_2:Dy is unstable when exposed to normal ambient levels of light. The spectral emission peaks are at 460 and 480 nm in the blue and 577 nm in the yellow region.

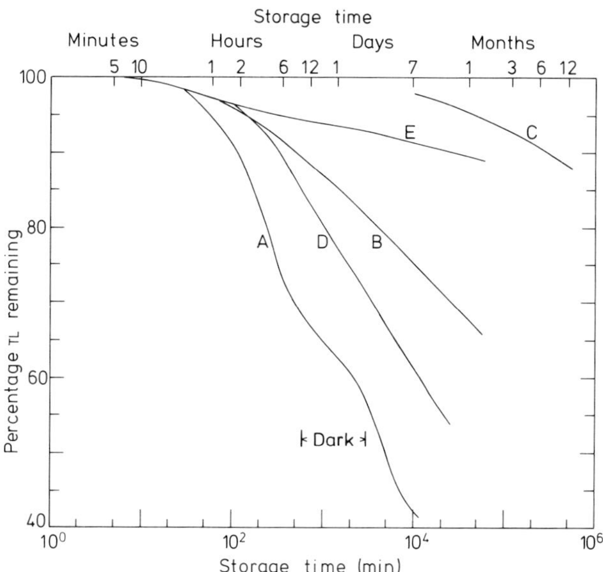

Figure 3.8 Thermal fading of CaF_2 phosphors. A and B, CaF_2:Dy 'light' and 'dark' fading respectively at normal ambient temperatures; C and D, CaF_2:Dy at normal ambient temperature and 50 °C respectively (Burgkhardt *et al* 1977); E, CaF_2:Mn at normal ambient temperature (Dekker 1976). (C and D, pre-read anneal of 20 min at 100 °C.)

Thulium-doped calcium fluoride phosphor was introduced by Lucas and Kapsar (1977). As with dysprosium-doped material the glow curve structure is somewhat complex, but there are dominant glow peaks at 150 °C and 250 °C. Each glow peak displays a different TL–absorbed dose response. The response of the 250 °C peak becomes markedly supralinear with increasing absorbed dose (times 3 at 10^3 Gy), while the 150 °C peak displays no supralinearity whatsoever. A marked increase in the sensitivity of the 250 °C peak compared with the 150 °C following irradiation with alpha radiation indicates that the material might prove useful for high LET, neutron and mixed-field dosimetry. The thermoluminescence spectral emission is in the blue, with peaks at 450 and 475 nm.

Natural fluorites, CaF_2 : Mn, CaF_2 : Dy and CaF_2 : Tm are commercially available.

Glow curve and TL–absorbed dose response. A summary of the glow peak structure and TL–absorbed dose response of calcium fluoride phosphors is presented in table 3.6.

Table 3.6 Summary of glow peaks and absorbed dose responses of calcium fluoride phosphors.

Activators	Glow peak temperature (°C)	Linear absorbed dose range (Gy)
Natural† Gd, Dy, Er, Tb, U	Variety of low-temperature peaks—260, 275	$<10^{-5}$–50‡ saturates at 10^3–10^4
Mn	260, high-temperature peaks	$<10^{-5}$–10 saturates at 10^4
Dy	120, 140, 200, 240 340, 400	$<10^{-5}$–10, 10^3§
Tm	150 and 250	$<10^{-5}$–10^2 (150 °C)

† Fluorites.
‡ Varies with sample of fluorite.
§ Linear response for 140 °C peak after pre-irradiation anneal at 600 °C for 2 h.

Relative photon energy response. The relative photon energy response of CaF_2 is shown in figure 3.4.

Fading. The low-temperature traps of the fluorites fade rapidly at normal ambient temperatures, but the dosimetry traps at 260 and 275 °C exhibit negligible thermal fading. The half-life of the 260 °C peak is approximately 2×10^5 years (Adam and Katriel 1971).

The 'dark' fading of CaF_2:Mn at normal ambient temperature is surprisingly high. The 'dark' fading of CaF_2:Dy is of the order of 10% after 24 h storage at normal ambient temperature, but can be reduced by a 20 min pre-read at 100 °C. The results of fading studies on CaF_2 phosphors are presented in figure 3.8.

3.6 Other Phosphors

3.6.1 Beryllium oxide

Beryllium oxide (BeO) is used extensively as an electrical insulating and refractory material. It is also used as a neutron moderator in nuclear reactors. Several different types of BeO are commercially available and certain ones have proved suitable for use as TL phosphors (Moore 1957, Tochilin *et al* 1969, Scarpa 1970a, b, Crase and Gammage 1975, Busuoli *et al* 1977), but the material has not achieved widespread use. BeO is tissue-equivalent, with a photon effective atomic number $Z_{\text{eff}} = 7.13$.

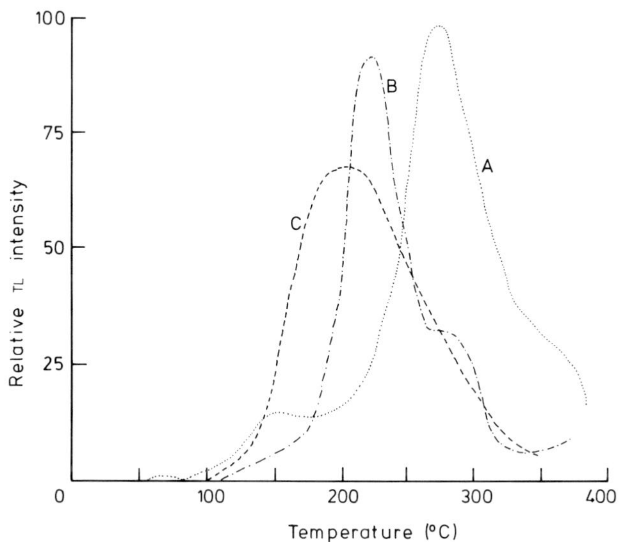

Figure 3.9 Glow curves for three different preparations of BeO. A, hot press quality; B, nuclear quality; C, slip cast (Scarpa 1970b).

Although the material is extremely toxic if inhaled as powder (see Appendix III), the sintered ceramic forms are safe provided they are handled with reasonable care.

The glow peak structure depends on the preparation parameters, the final physical form of the dosemeter, the magnitude of absorbed dose and the relative proportions of the commonly found impurities, such as silicon, sodium, fluorine, sulphur and carbon. Glow curves obtained from three different preparations of BeO are illustrated in figure 3.9. Depending on the impurities, form of dosemeter, and LET of the radiation, the TL–absorbed dose response has a threshold of $< 10^{-4}$ Gy, is linear up to between 0.5 and 5 Gy, and saturates at approximately 5×10^3 Gy.

The TL emission spectrum of BeO extends from the blue region of the visible spectrum into the ultraviolet, down to 200 nm wavelength (Mandeville and Albrecht 1954, Spurny and Hruska 1976). A higher detection sensitivity can be achieved using a photomultiplier tube fitted with a quartz window and ultraviolet transmitting optics.

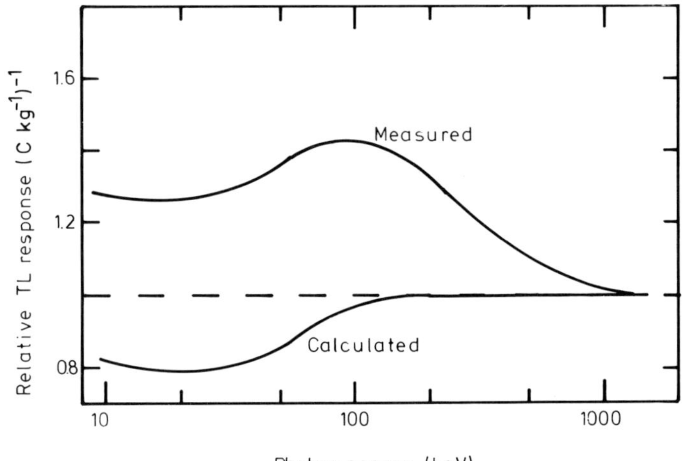

Figure 3.10 Calculated and measured relative photon energy responses of BeO phosphor. A, experimental; B, theoretical. (Tochilin *et al* 1969, reprinted with the permission of Pergamon Press Ltd, Oxford.)

While the theoretical relative photon energy response of BeO matches that of air very closely, experimental data differ considerably from theory, as illustrated in figure 3.10.

The 'dark' thermal fading of BeO is low at normal ambient temperature, but light-induced fading is very high: 50% after 30–60 min exposure to normal ambient levels of fluorescent lighting (Scarpa 1970b, Crase and Gammage 1975). The intrinsic response to ultraviolet radiation is relatively low and, like LiF, the absorbed dose can be re-assessed by phototransferred thermoluminescence.

3.6.2 Aluminium oxide

Aluminium oxide (Al_2O_3), like BeO, is a refractory material and is used for many industrial applications. Its various forms contain many different impurities, including calcium, chromium, titanium, nickel, magnesium, sodium and iron, etc. The very hard precious stones, ruby and sapphire, are particular forms of Al_2O_3. The TL properties of Al_2O_3 in its various forms have been widely investigated (Rieke and Daniels 1957, McDougall and Rudin 1970, Mehta and Sengupta 1976). The material is reasonably tissue-equivalent down to about 150 keV, it has a twofold over-response at 100 keV, and a threefold maximum over-response at lower photon energies.

The various commercial grades of Al_2O_3 contain many different impurities, some of which are listed above, and many different crystalline phases depending on the thermal treatment. Both of these factors critically affect the glow peak structure and the TL sensitivity of the phosphors. For example, Reike and Daniels (1957) identified four glow peaks at approximately 103, 123, 164 and 236 °C. The 236 °C peak is particularly interesting as it is associated with the presence of sodium and is not present unless the phosphor is annealed at a temperature above 1000 °C. This anneal results in the transformation of the lattice to a single-crystalline form.

A similar glow peak structure has been observed in industrial corundum, another commercially available form of Al_2O_3, which displays glow peaks at 50, 125, 150, 325, 475 and 625 °C. The relative sensitivity of this phosphor depends on the presence of silicon and titanium (Mehta and Sengupta 1976). Using a detector with an appropriate spectral response, minimum detectable absorbed doses ranging from a few tens of μGy to 1 mGy have been reported for the most sensitive forms of Al_2O_3. The TL–absorbed dose responses are linear up to between 1 and 30 Gy, with saturation at approximately 10^4 Gy.

TL emission spectra also vary considerably with the type of Al_2O_3. The spectral emission for ruby is at 700 nm in the red region, while an emission peak at 420 nm in the blue is associated with those phosphors containing titanium.

The low-temperature peaks of Al_2O_3 fade rapidly at normal ambient temperature but the 236 °C and higher-temperature peaks are very stable and exhibit negligible thermal fading. However, light-induced fading can be high for some phosphor preparations.

3.6.3 Magnesium orthosilicate

Magnesium orthosilicate (Mg_2SiO_4) occurs in the natural mineral forms of olivine and forsterite. Hashizume *et al* (1971) were the first to manufacture a Mg_2SiO_4 TL phosphor containing the rare earth terbium as a dopant. The characteristics of the phosphor depend on the preparation techniques used, and in particular on the temperature of the melting process.

The TL sensitivity is between 40 and 100 times that of LiF:Mg:Ti phosphor using a standard photomuliplier detector (Hashizume *et al* 1971, Bhasin *et al* 1976). The TL–absorbed dose response is linear from a few tens of μGy to 4 Gy. Thereafter it is supralinear to $>$ 200 Gy, with saturation at between 500 and 10^4 Gy, depending on the phosphor preparation.

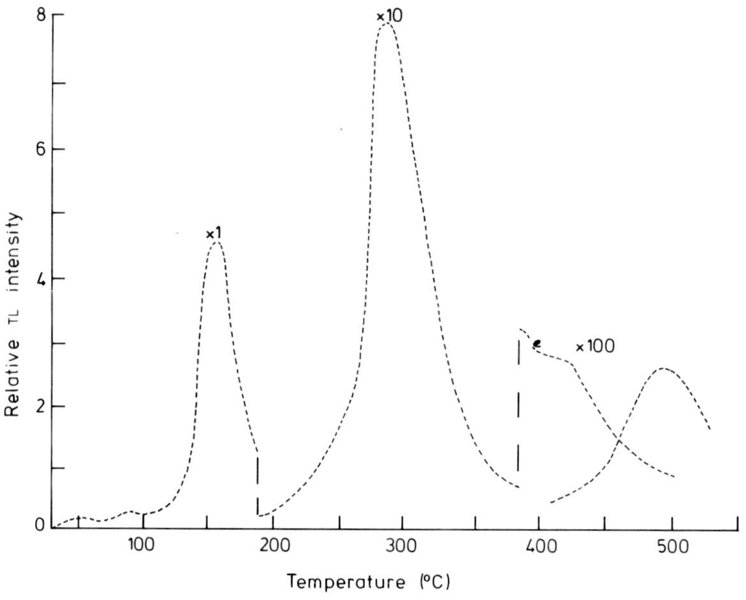

Figure 3.11 Glow curve for $MgSiO_4$ phosphor prepared at a high temperature (2750 °C). (Bhasin *et al* 1976, reprinted with the permission of Pergamon Press Ltd, Oxford.)

Some forms of the phosphor display a broad dosimetry peak at around 200 °C (Hashizume *et al* 1971, Nakajima *et al* 1971, Jun and Becker 1975, Lakshmanan and Vohra 1979), while one form prepared at a very high temperature (2750 °C) has a 300 °C dosimetry peak with a number of other relatively minor ones (Bhasin *et al* 1976), as shown in figure 3.11.

In view of its high sensitivity, it is not surprising to find that the main spectral emission peak of Mg_2SiO_4:Tb matches the spectral detection peak of the standard TL reader photomultiplier detector, i.e. 380–400 nm.

The photon energy response shows an enhancement factor of between 4 and 5.5 to 30–40 keV photons compared with ^{60}Co radiation. Thermal fading at normal ambient temperature is negligible; indeed, one form of the phosphor (Bhasin *et al* 1976) displays only approximately 17% fading after 2 h post-irradiation storage at 200 °C. Optically induced fading is high, however, with radiation of wavelengths 370 and 500 nm producing the greatest effect.

Spurious photoluminescence is also high (Nakajima 1972). The phosphor must be adequately shielded from natural and artificial illumination. A simple pre-irradiation anneal of up to 3 h at 500 °C ensures reproducibility of absorbed dose response following re-use of the phosphor.

3.6.4 MgB_4O_7:Dy:Tm

This recently produced phosphor (Prokic 1980) is available in the form of physically robust sintered discs 9, 6, 5 and 3 mm in diameter. It has a main glow peak at 210 °C (the same as LiF) and its main advantages appear to be (i) a TL sensitivity between five and ten times that of LiF (the useful measurement range of absorbed dose is 10^{-5}–10^{3} Gy); (ii) fading similar to that of $Li_2B_4O_7$:Mn; and (iii) a simple annealing procedure (10 min at 300 °C) which is required only following high levels of exposure. Initial results from independent studies on the material indicate that the reproducibility of absorbed dose measurements is very good ($< 0.5\%$ at 10^{-3} Gy) and that the photon energy response is similar to that of LiF (35–40% over-response to photons of energy < 70 keV). Unlike previous forms of borate phosphor, the material apparently does not exhibit any adverse effects resulting from exposure to high humidity.

3.6.5 $CaSO_4$:Dy:Tm

This material, also developed by Prokic (1980), is available in the same forms as MgB_4O_7 detailed above. The dosimetry glow peak is at 230 °C and the useful measurement range of absorbed dose is 10^{-6}–10^{3} Gy. The reported thermal fading at normal ambient temperatures (< 30 °C) is approximately 5% per year. As with the MgB_4O_7 phosphor, the annealing requirements are very simple—1 h at 300 °C following heavy irradiation.

3.7 Physical Forms of Dosemeters

Thermoluminescence dosemeters are available in two general forms:

either as loose phosphor powder or as solid dosemeters. The solid dosemeters may be composed entirely of phosphor as single crystals and polycrystalline extrusions, or as a homogeneous composite of the phosphor powder and some binding material such as PTFE (figure 3.12).

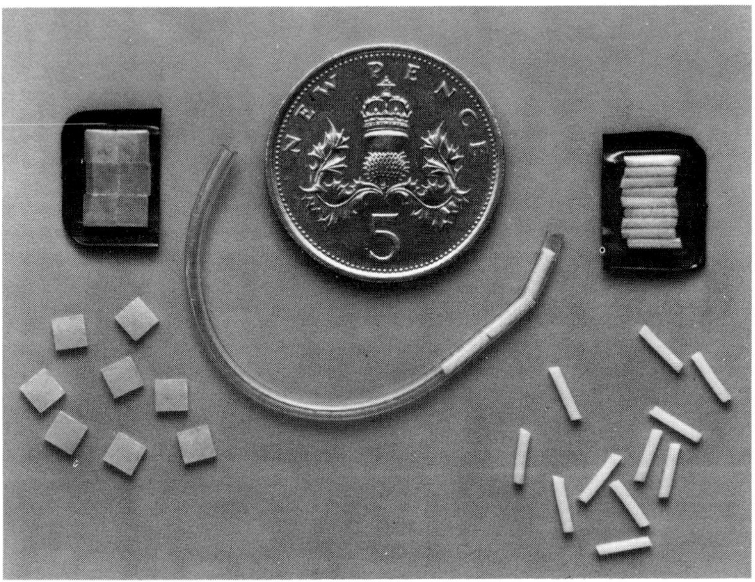

Figure 3.12 Solid forms of TL phosphor. *Left,* TLD 700 extruded-ribbon dosemeters; *right,* LiF:PTFE micro-rod dosemeters; *centre,* micro-rods sealed in catheter tubing. (Photograph by courtesy of the National Radiological Protection Board, Harwell.)

The characteristics of the 100% phosphor dosemeters may be considerably different from those of the composites. Within limits, the thermoluminescence sensitivity of a dosemeter is proportional to the mass of phosphor present and, because the maximum loading fraction of phosphor consistent with good mechanical properties is about 30% wt/wt, these dosemeters are less sensitive. Their sensitivity may be further affected by self-absorption of the thermoluminescence signal. The required sensitivity will vary greatly according to the type of absorbed dose measurement; for example, LiF:PTFE micro-rod dosemeters containing 4% wt/wt LiF are suitable for many radiotherapy measurements where individual fractional absorbed doses of a few Gy are to be measured, but are of no use in personal monitoring. The lowest detectable absorbed dose is not solely a function of thermoluminescence sensitivity, however, but depends on the luminescence background signal inherent in the dosemeter and generated during readout. This background may be due to a number of effects, many of which are related to the surface of the dosemeter and its interaction with the environment (e.g. triboluminesence, chemiluminescence, etc), and are generally much less for solid form dosemeters than for loose powder. Weighted against this is the increase in background signal resulting from volume luminescence effects in the binder material of composite dosemeters.

The measurement precision attainable with any dosemeter may be considered in two ways: either the reproducibility of thermoluminescence signals from one particular dosemeter which is identically re-used many times, or the spread of sensitivities of many nominally identical dosemeters used once. As has been pointed out by Robertson (1975) the attainable precision is dependent on the magnitude of the measured absorbed dose. At levels of absorbed dose whose equivalent thermoluminescence signal is much greater than the luminescence background the important parameters affecting precision are those associated with the manufacture of the dosemeter and with its proper handling and use. At low levels of absorbed dose where the luminescence background is of the same order of magnitude as the measurement signal one must additionally consider the variability and resultant uncertainty in the luminescence background. The cost of production and/or selection of large numbers of identically sensitive dosemeters is high and often the user would do better to use a batch of nominally identical dosemeters and calibrate each dosemeter individually. The useful technique of sensitivity pairing of dosemeters can then be applied. Repeat calibration is necessary to ensure subsequent precision.

3.7.1 Powder dosemeters

Single crystals of TL material, while providing a very sensitive form of dosemeter, display a very non-uniform distribution of TL sensitivities and it has been found that by grinding up many such crystals and thoroughly mixing the resultant powder, a reasonably uniform TL sensitivity can be achieved. The sensitivities of powder phosphors, and in some cases the glow peak structures, have been shown to be strongly dependent on grain size, most notably in LiF:Mg:Ti (Zanelli 1968, Robertson 1975). Therefore powder is usually sieved to a range of grain sizes which, in the case of commercial LiF:Mg:Ti, is 75–200 μm. Grains of this size are free-running and therefore easily dispensed and they have an adequate TL sensitivity. Smaller grains tend to adhere to each other and to their container, and larger grains may not provide sufficiently uniform sensitivity over a number of small-volume samples.

Because powder acts essentially like a fluid it adopts the shape of the holder in which it is placed, thus providing a very useful degree of flexibility in the choice of dosemeter size and shape. Batches of powder can be conveniently annealed in bulk at high temperatures.

The dispensing of powder samples is usually carried out using vibrating volume dispensers which, when new, can generally dispense equal volumes of powder with a reproducibility of better than 1% although, as pointed out by Karzmark *et al* (1965), some dispensers may wear over a period of a year or two resulting in uncertainties of 2–3%. The distribution of grain sizes of the powder may also change due to the abrasion received during repeated dispensings. For precision dose assessment accurate weighing of the powder (± 0.1 mg) is necessary.

As detailed in § 7.3, weighing should be done immediately prior to readout to avoid undetected loss of powder either during irradiation or readout. The uncertainty in the uniformity of distribution and positioning of the powder on the reader heating tray is a further limitation on the precision obtainable with powder samples. For example, too thick a sample may result in inefficient heating, thermal gradients, self-absorption and scatter of the TL signal, and exposure of an area of the heating tray to the photomultiplier detector.

In powder samples the ratio of surface area to volume is high compared with solid form dosemeters and therefore surface-related effects such as tribothermoluminescence and chemiluminescence assume more importance. These effects can be greatly reduced, however, by using the standard readout technique of inert gas flow, usually oxygen-free nitrogen. Powder dosemeters are now also available in bulb form which eliminates these effects.

3.7.2 Extruded and hot pressed dosemeters

The extruded and hot pressed forms of dosemeter are available in two main geometries: the ribbon (commonly called the chip) and the rod (commonly called the micro-rod), both of which are produced by compression of the normal phosphor ingredients at an elevated temperature (e.g. for LiF approximately 3500 kg cm^{-2}). For extruded dosemeters the fused polycrystalline material is extruded through an appropriately shaped die, cut and polished. The hot pressed forms have a higher sensitivity than the extruded types, although the useful TL characteristics of both are similar to those of loose powder. A variety of sizes are available but the currently most popular ones are $3 \times 3 \times 0.9$ mm chips and $1 \times 1 \times 6$ mm micro-rods. Like powder they can be annealed at high temperature (e.g. LiF at 400 °C), and are easily handled and washed. They have found widespread application in the fields of personal, clinical and environmental dosimetry.

3.7.3 PTFE-based dosemeters

By compressing and heating a homogeneous mixture of fine-grain phosphor powder (for LiF typically 10 μm) and PTFE powder to a temperature above the softening temperature of PTFE (327 °C) in a mould, it is possible to form an intimate matrix of phosphor and PTFE. By varying the loading fraction of the phosphor powder the sensitivity of the dosemeters can be changed. PTFE-based dosemeters are available with different phosphors and in a variety of geometries. Discs are available with diameters of 2–13 mm and thicknesses of 0.02–0.5 mm. Micro-rods have also been produced in various lengths and 1 mm diameter. In the UK 13 mm diameter × 0.4 and 0.2 mm thick LiF:PTFE discs (30% wt/wt LiF) have found widespread use for personal dosimetry applications, and micro-rods 1×6 mm (4% wt/wt LiF) have been used for clinical applications, in particular for radiotherapy absorbed dose measurement.

PTFE-based dosemeters cannot be annealed in bulk at temperatures > 300 °C, and one manufacturer recommends that the cooling rate from this temperature should not exceed 36 °C h^{-1} to ensure adequate stress relief of the PTFE and to achieve uniform optical density. However, using a reader anneal of 300 °C for 15 s followed by a low-temperature anneal of 16–24 h at 80 °C, LiF:PTFE disc dosemeters have been successfully re-used for personal dosimetry over many cycles (Marshall *et al* 1971). Spurious thermoluminescence can be induced in PTFE-based dosemeters by exposure to ultraviolet radiation; e.g. LiF:PTFE discs exposed to direct sunlight can receive an induced TL signal equivalent to up to 1.5 mGy. A prompt ultraviolet-induced phosphorescence has also been noted by Robertson (1975). Both effects may be avoided by packaging the dosemeters in an opaque envelope. The attraction of dust by electrostatic charges on the disc is a further problem associated with the use of PTFE dosemeters.

Table 3.7 Characteristics of some

Phosphor characteristics	LiF	$Li_2B_4O_7$:Mn	CaF_2:Nat
Density (g cm^{-3})	2.64	2.3	3.18
Photon effective atomic number	8.2	7.4	16.3
Ratio of TL response 30 keV/^{60}Co	1.3	0.9	13–15
Main glow peak temperature (°C)†	190–210	200–220	200, 275
Spectral emission peaks (nm)	400	600	380
Fading of dosimetry peak at normal ambient temperature	5% in 3–12 months‡	5–10% in 3 months	<3% in 9 months
Useful range of absorbed dose (Gy) §	5×10^{-5}–10^{3}	10^{-4}–10^{4}	$<10^{-5}$–10^{2}
Chemical stability	Good	Hygroscopic	Good
Toxicity‖	High if ingested	High if ingested	Low
Physical forms	Crystals, powder, bulbs, chips micro-rods, PTFE discs, micro-rods, tape	Powder, bulbs, chips, PTFE-based micro-rods	Crystals
Principal applications	Personal, Accident and radiotherapy dosimetry	Radiotherapy and diagnostic radiology dosimetry	Low absorbed dose measurement

† Depends on heating rate and preparation.
‡ Depends on pre- and post-irradiation annealing.
§ Depends on sample form.
‖ See Appendix III.

commercially available TL phosphors.

CaF_2:Mn	CaF_2:Dy	$CaSO_4$:Mn	$CaSO_4$:Tm	$CaSO_4$:Dy	BeO
3.18	3.18	2.61	2.61	2.61	3.01
16.3	16.3	15.3	15.3	15.3	7.1
13–15	13–15	10–12	10–12	10–12	1.0
260	200, 240	110	220, 250	220, 250	180, 220
500	480, 577	500	452	478, 571	330
15% in 2–4 weeks (10% in first 24 h)‡	25% in 4 weeks (10% in first 24 h)‡	35% in 24 h	6% in 6 months	6% in 6 months	≯5% in 5 months
10^{-5}–2×10^{3}	10^{-6}–10^{3}	10^{-7}–10^{2}	10^{-6}–10^{2}	10^{-6}–10^{3}	10^{-4}–10^{3}
Good	Good	Good	Good	Good	Good
Low	Low	Low	Low	Low	Powder—high if inhaled
Powder, bulbs, chips, micro-rods	Crystals, powder, bulbs, chips	Powder	Crystals	Powder, discs, chips	Ceramic
High or low absorbed dose measurement	High or low absorbed dose measurement	Environmental and short-term dosimetry	Environmental dosimetry	Environmental dosimetry	Personal dosimetry

3.7.4. Silicone-embedded dosemeters

Dosemeters have been produced by incorporating the phosphor powder in silicone rubber, but two main problems have been encountered in their use. Some of the silicone rubbers used have proved to be thermoluminescent when exposed to ultraviolet radiation (even to fluorescent room lighting) and they cannot be heated above 280 °C. The latter condition is particularly important with regard to annealing for re-use. They have therefore not found widespread application.

3.7.5 Sintered dosemeters

Sintered dosemeters are formed by compression of the phosphor powder in a press mould followed by a high-temperature sinter. A major limitation of some of these dosemeters is their physical fragility—particularly in large-diameter thin form—and although sintered LiF, BeO, $Li_2B_4O_7$, Al_2O_3, MgB_4O_7 and $CaSO_4$ have all been produced, they have not found widespread use.

3.7.6 Commercially available phosphors

A résumé of the TL characteristics of readily available commercial forms of phosphor discussed above is presented in table 3.7.

4 Applications of Thermoluminescence Dosimetry in Medicine

4.1 The Need for Clinical Absorbed Dose Measurement

There are two important areas of absorbed dose measurement in medicine: (i) absorbed dose measurement in radiotherapy, and (ii) absorbed dose measurement in diagnostic radiology.

4.1.1 Radiotherapy measurements

As the objective of radiotherapy is to deliver a prescribed absorbed dose to the cancerous tissue (the target volume) over a prescribed time interval, and to minimise the absorbed dose to surrounding tissue and other organs, it is important to attempt to assess this absorbed dose as precisely as possible.

The difficulties of accurately predicting treatment absorbed doses by calculation, have in the past led to the development of *in vivo* measurement techniques. While entrance and exit exposures could be measured using films and conventional ionisation chambers, intra-cavitary measurements were limited by the degree of miniaturisation of chambers which could be practically achieved (Sievert 1934). These so-called 'condenser' miniature chambers proved very useful for certain types of measurement including, for example, the measurement of exposure to the oesophagus during rotation therapy (Liden 1948). Typically chambers and lead markers were sealed in rubber or plastic sheaths and positioned by x-ray film exposure. However, their size (typically 20×5 mm) precluded their use for many other types of intra-cavitary measurement.

When Daniels first developed thermoluminescence as a practical method of ionising radiation dosimetry, it was realised almost immediately that it could be applied in the field of clinical measurement. Brucer used some of Daniels' single crystals of Harshaw LiF to make internal *in vivo* measurements in cancer patients injected with radioactive material (Daniels *et al* 1953). The procedure was simple: the patients swallowed the crystals which were recovered and read out several days later. Exposures of up to 60 R (1.55×10^{-2} C kg^{-1}) were measured by this method.

The arrangement of radiotherapy treatment fields may be done using a combination of calculations involving standard geometrical set-ups, and depth dose and transverse dose measurements in phantoms. The final check on the treatment absorbed dose delivered to the patient can be carried out by *in vivo* dosimetry. Similarly spared (shielded) organ absorbed doses, which should be kept to the minimum consistent with effective treatment, can be measured.

Thermoluminescence dosimetry has proved a useful technique for a variety of purposes in radiotherapy, including measurements of therapy machine output, beam uniformity checks, and the measurement of absorbed doses in phantoms and *in vivo* for both internally and externally applied fields. TL dosemeters give high precision, rapid retrieval of information using on-site readers, good environmental stability, good water or tissue equivalence, and a wide range of sensitivities. The last characteristic is particularly important for parallel *in vivo* measurements of radiotherapeutic absorbed dose and the relatively low absorbed dose to spared organs. Because of their small size TL dosemeters also give good spatial resolution, which is of particular value in many radiotherapy techniques where high absorbed dose gradients may be used. In biology too, TL dosemeters can be used to measure the absorbed dose to experimental animals both from externally applied fields and from ingested radioisotopes.

4.1.2 Diagnostic radiology measurements

The collective dose resulting from medical exposure has been estimated to represent the largest single man-made contribution to both the somatic and the genetically significant dose equivalent (GSD) to the population of the United Kingdom (as illustrated in figure 4.1), representing some 31% of the total somatic and 9% of the total GSD, and 95 and 85% respectively of the man-made contributions. These compare with approximately 0.4% of each total for occupational exposure to ionising radiations and 0.2% from radioactive waste disposal. In other developed countries the estimated figures are similar. By far the largest contribution is from diagnostic radiology, estimated as ten times the sum of the contributions from nuclear medicine and radiotherapy. It is therefore desirable to reduce the exposure of patients to diagnostic x-rays to the minimum consistent with good diagnosis. New radiodiagnostic techniques have been introduced, some involving the use of complex machines such as the head and body scanners (computerised axial tomography—CAT). Thermoluminescence dosimetry has proved to be

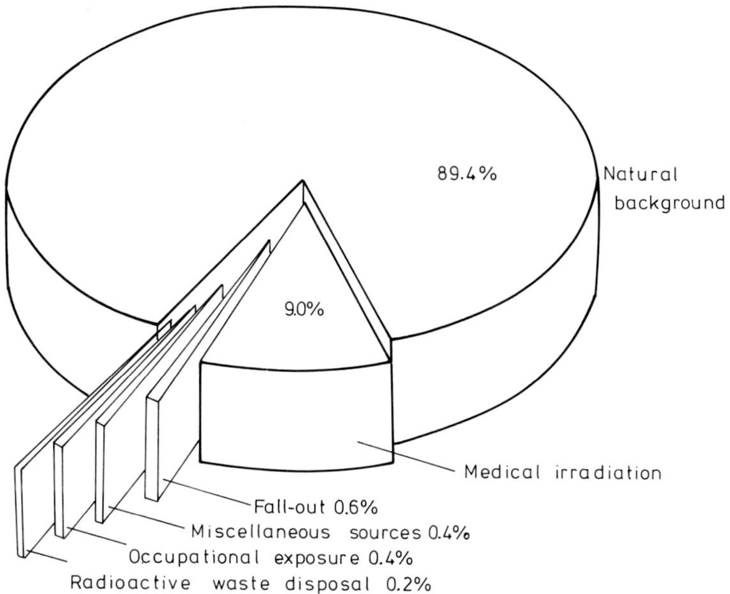

Figure 4.1 Annual genetically significant dose equivalent to the UK population. (Taylor and Webb 1978, reprinted with the permission of the National Radiological Protection Board, Harwell.)

a useful method in the comparison of patient absorbed dose from these new techniques and from more traditional methods (for CAT measurements see, e.g. Horsley and Peters 1976, Wall *et al* 1979b; for dental pantomography measurements see Wall *et al* 1979a).

The importance of diagnostic absorbed dose measurements lies in:

(1) improving the design of equipment to reduce patient absorbed dose;

(2) improving radiographers' techniques in the use of equipment to reduce patient absorbed dose; and

(3) providing a measurement data base for epidemiological analysis of population radiation absorbed dose from diagnostic radiology.

Thermoluminescence phosphors such as LiF:Mg:Ti and particularly $Li_2B_4O_7$:Mn have been used successfully in such measurements. They have three main advantages over ionisation chambers for this type of measurement:

(1) they are small and unobtrusive;
(2) they are radio-transparent to most x-radiations; and
(3) they are independent of leads and are easily attached to the patient.

4.2 Dosemeters for Clinical Use

A number of factors must be considered in the choice of the material and the form of TL dosemeter for clinical dosimetry applications. The most important factors are:

(1) estimated absorbed dose range of the intended measurements;
(2) estimated photon energy or LET of the radiation;
(3) immediate environmental conditions around the dosemeter; and
(4) degree of spatial resolution required.

4.2.1 Absorbed dose range

The sensitivity of any TL dosemeter is proportional to the mass of active phosphor present, within limits imposed by geometrical and thermal considerations relevant to the readout system. Dosemeters which contain only active material, such as powders, extruded ribbons and rods, have a much higher TL sensitivity than dosemeters consisting of phosphor embedded in a matrix of binder material, such as PTFE-based discs, tape and micro-rods.

The range of absorbed doses encountered in clinical and biological irradiations is very large. In diagnostic radiology absorbed doses may range from less than 10^{-5} to 10^{-1} Gy to the gonads and from approximately 10^{-4} to about 1 Gy to the skin. Hence at the lower end of the diagnostic radiology range there is a need for a sensitive form of dosemeter, e.g. powder, extruded ribbon or extruded micro-rod. In therapy dosimetry, however, a single treatment fraction of absorbed dose may be several Gy, and in many animal or cell irradiations tens of Gy may be required. Less sensitive forms of dosemeter may be used, especially those incorporating the phosphor powder in PTFE as discs, tape and micro-rods.

The measurement of high absorbed doses will often necessitate the use of the dosemeter above the linear TL–absorbed dose response region;

under these conditions a supralinearity calibration will have to be carried out. Often in radiotherapy, simultaneous measurements of treatment absorbed dose and the much smaller absorbed dose in the penumbra or umbra of shielding may be required. In such cases two types of dosemeter, such as extruded rods and PTFE-based rods, may be used in parallel. This particular combination of dosemeters should provide good spatial resolution if appropriately oriented with respect to the absorbed dose gradient.

Sensitisation is a problem associated with the re-use of dosemeters after the measurement of very high absorbed doses. This is of particular importance with the use of PTFE-based dosemeters as they cannot be adequately re-annealed at temperatures much in excess of 300 °C.

4.2.2 Photon energy range—LET of radiation

Modern routine radiotherapy and radiodiagnostic techniques use a wide range of photon energies of approximately 10–200 keV conventional x-ray machines, ^{137}Cs (0.67 MeV) and ^{60}Co (1.25 MeV) teletherapy units, and accelerators producing megavoltage photons and high-energy electrons.

As discussed in Chapter 3, the total TL emitted by an irradiated phosphor is proportional to the total radiation energy absorbed by it. In tissue the most important absorption process in the photon energy range approximately 20 keV to 10 MeV is the absorption of Compton scatter electrons. For low atomic number elements such as lithium, boron, oxygen, fluorine, etc, and for photon energies up to about 15 keV, the photoelectric effect is dominant. Thereafter, up to 10 MeV, Compton scatter is most important. For elements of high atomic number, which TL dopants often are, the photoelectric effect is dominant up to several hundred keV. The theoretical photon energy responses of a number of commonly used phosphors are illustrated in figure 3.4. The advantage of using TL materials consisting mainly of low atomic number atoms such as Li, B, O, F, etc, with only a few higher atomic number atoms such as Mn, Mg, Ti, etc, is immediately obvious because of their good approximation to tissue and air. This is particularly true of lithium borate based phosphors.

In order to evaluate the absorbed dose to a phantom or patient using TL dosemeters, it is essential to know the relative energy responses of the dosemeters throughout the range of energies used. The primary response calibration of dosemeters is usually carried out using a ^{60}Co source (1.25 MeV) and the response at all other energies and for all other radiations is expressed as a multiple or fraction of this. For clinical applications the response is most usefully expressed as the TL per unit absorbed dose in tissue or water. This will be a function of energy and the physical form of the dosemeter, and this response factor can be obtained by measurement, in principle using the methods described in § 7.4.

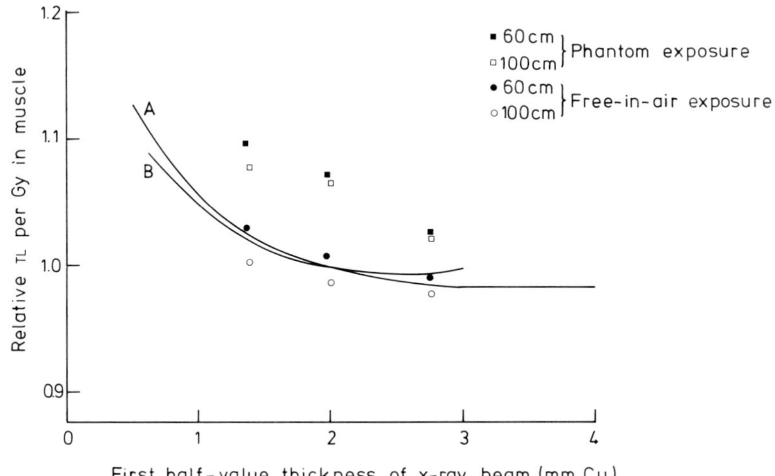

Figure 4.2 Relative TL response of LiF:Mg:Ti per Gy in muscle. A, theoretical; B, obtained experimentally under standardised exposure conditions (exposure distance 141 cm). Experimental points obtained for free-in-air and in phantom exposure at 60 and 100 cm, as indicated (Puite and Crebolder 1974).

The beam quality has to be known. Photon beam quality determination, although in principle relatively straightforward in free air, is difficult to determine uniquely in water or solid phantoms because of two effects:

(1) the contribution from lower-energy scattered radiation from the phantom and external shielding material, e.g. the applicator cone at short focus to skin distances (FSD); and

(2) the effective 'hardening' of the beam with increasing depth in the water or solid phantom.

These two effects are illustrated in figure 4.2, which shows the results of energy dependence measurements using LiF:Mg:Ti dosemeters in a mouse phantom (Puite and Crebolder 1974). In addition, by comparing the results of measurements of the ratio of the differing and energy-dependent responses of CaF_2:Mn and LiF:Mg:Ti powder dosemeters exposed in the phantom with others exposed in free air, effective half-value layers (HVL) within the phantom can be obtained.

The TL–absorbed dose (water and polystyrene) response of LiF:Mg:Ti to high-energy radiations, of which the ones of principal interest in clinical applications are megavoltage photons and high-energy electrons, has been widely investigated and reported in the literature.

The results have often been inconsistent. Some investigators have measured an approximate 10% reduction in response for high-energy radiations compared with that from ^{60}Co, while others have not found any reduction. Much discussion has revolved around the application of various general cavity theories to attempt to explain the observed effects and reconcile the differences. For useful reviews and entry into the literature of this subject the reader is recommended to the works of Paliwal and Almond (1975) and Rudén and Bengtsson (1977).

The relative energy responses of a number of different physical forms of LiF:Mg:Ti, and powdered $Li_2B_4O_7$:Mn and CaF_2:Mn dosemeters for a range of photon and electron energies of particular relevance to clinical use are presented in table 4.1.

4.2.3 Environmental factors

Absorbed dose range and radiation energy considerations apart, environmental factors such as temperature, humidity, contact with body fluids, insertion into catheters, sterilisation, etc, will influence the choice of dosemeter form or packaging. If dosemeters are not protected from their environment the result is often low precision and sometimes gross error in absorbed dose measurements.

4.2.3.1 Temperature and humidity. During exposure under clinical conditions, dosemeters may come into contact with heat (human body core temperature is 37 °C) and/or high-humidity environments. If implanted or introduced into body cavities, they may come into contact with body fluids; thermal fading apart, some phosphors (especially in powder form) have been shown to be affected by humidity. This is particularly true of $Li_2B_4O_7$:Mn, as illustrated in figure 4.3.

4.2.3.2 Other agents. If solid form dosemeters are attached directly onto the skin using adhesive tape, care should be taken to remove all traces of adhesive from the dosemeter before readout. Adhesives often

Table 4.1 Measured values of TL per Gy (in water) for TL dosemeters relative to ^{60}Co.

Dosemeter radiation	0.1 mm LiF:PTFE discs†	0.4–0.5 mm LiF:PTFE discs†	LiF extruded ribbons†	LiF:PTFE rods†	LiF rods†	LiF powder‡	CaF_2 powder‡	$Li_2B_4O_7$:Mn powder‡
Photons (keV)								
10	1.2	0.95	0.78	—	—	—	—	—
17	1.4	1.31	1.36	—	—	—	—	—
23	1.45	1.38	1.43	—	—	—	—	—
37	1.43	1.42	1.45	—	—	—	—	—
51	1.36	1.35	1.38	—	—	—	—	—
81	1.15	1.15	1.17	—	—	—	—	—
97	1.12	1.12	1.13	—	—	—	—	—
^{60}Co(MeV)								
1.25	1.00	1.00	1.00	1.00	1.00	1.00	1.00	1.00
^{60}Co(MV)								
6	0.96	0.94	0.97	0.97	0.97	—	—	—
18.5	—	—	—	—	—	0.94	0.96	—
22	—	—	—	—	—	0.93	0.96	0.98
35	—	—	—	—	0.90§	—	—	—
42	0.96	0.93	0.96	0.97	—	—	—	—
50	—	—	—	—	0.91§	—	—	—
65	—	—	—	—	0.91§	—	—	—

Electrons (MeV)								
4.3	0.93	0.90	0.90	0.92	0.92	—	—	—
5	—	—	0.95‖	—	—	—	—	—
6	—	—	—	—	—	0.94	—	—
7.4	0.93	0.91	0.91	0.94	0.94	—	—	—
9	—	—	—	—	—	—	0.94	0.97
9.8	0.93	0.91	0.91	0.94	0.94	—	—	—
10.0	—	—	0.90‖	—	—	—	—	—
11.6	0.93	0.91	0.91	0.94	0.94	—	—	—
12	—	—	—	—	—	0.93	0.98	0.96
14.3	0.94	0.91	0.92	0.95	0.95	—	—	—
15	—	—	—	—	—	0.92	0.97	0.96
18	—	—	—	—	—	0.89	0.97	0.97
19.4	0.96	0.92	0.93	0.96	0.96	—	—	—
20	—	—	0.92‖	—	—	—	—	—
28.2	0.96	0.92	0.94	0.96	0.96	—	—	—
30	—	—	0.94‖	—	—	—	—	—
35	—	—	0.95‖	—	—	—	—	—
39.2	0.98	0.93	0.95	0.97	0.97	—	—	—
40	—	—	0.92‖	—	—	—	—	—

† Rudén (1976).
‡ Almond and McCray (1970).
§ Turner and Anderson (1973).
‖ Bistrović *et al* (1976), Lucite phantom.

exhibit thermoluminescence following exposure to visible light and/or ultraviolet radiation. The simplest way of avoiding these effects is to seal the dosemeters inside thin protective envelopes (e.g. polythene). Should dosemeters become contaminated they may be cleaned using the methods outlined in § 7.3 or those specifically recommended by the manufacturer.

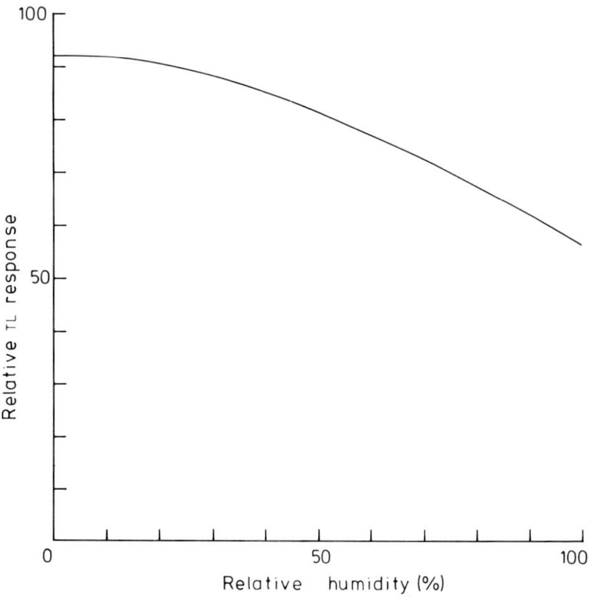

Figure 4.3 Effects of temperature and humidity on the TL response of $Li_2B_4O_7$:Mn, 38 days pre-irradiation storage at 40°C (Mason *et al* 1974).

4.2.3.3 Sterilising of dosemeters. On occasion, a clinician or biologist will require TLDs to be sterilised. The three common methods of sterilising: autoclaving, chemical sterilising and 254 nm ultraviolet radiation can all have gross effects on the inherent TL sensitivity of the dosemeter or may induce spurious luminescence. In general, provided the phosphor is effectively sealed in a protective envelope or catheter which is opaque to 254 nm ultraviolet radiation, either chemical or ultraviolet sterilising at normal ambient temperature is recommended. Normal ambient temperature is emphasised because the effects of elevated temperatures on the normal TL sensitivity of phosphors—especially LiF:Mg:Ti—can be significant, as discussed in Chapter 2. The effects of autoclaving can be particularly severe, as shown in table 4.2.

Table 4.2 Effects of temperature and humidity on the sensitivity of TL dosemeters.

Dosemeter	Treatment		TL per Gy (% standard deviation)	TL as % of controls†
LiF extruded ribbon	120°C‡	Wet	1132 (3%)	75
		Dry	1151 (2%)	75
LiF:PTFE micro-rods	120°C	Wet	20.1 (4%)	76
		Dry	19.7 (4%)	74
$Li_2B_4O_7$:Mn powder	120°C	Wet	9027 (8%)	50–60
		Dry	16115 (5%)	97–100

† Controls stored for equivalent time at 20°C, <20% relative humidity.

‡ Some materials are claimed autoclavable at temperatures up to 120°C (standard autoclave temperature, 132°C).

4.2.4 Spatial resolution

Good spatial resolution of absorbed dose measurement is generally useful, and in the measurement of high absorbed dose gradients it is essential. TL dosemeters are available in many shapes. Powder acts like a fluid and will adopt the shape of its container. The micro-rod and extruded-ribbon dosemeters are so small ($1 \times 1 \times 6$ mm and $3.2 \times 3.2 \times 0.9$ mm, respectively), that the effective size of the dosemeter is often limited in practice by the requirement to have adequate build-up to ensure electronic equilibrium.

The usefulness of TL dosemeters for high spatial resolution measurements is illustrated in figure 4.4 (Scarpa *et al* 1979). The radiation and shielding configuration shown is one used to irradiate only one lung of a mouse, and the absorbed dose profile was obtained using individually calibrated LiF:Mg:Ti extruded-ribbon dosemeters contained in a Perspex mouse phantom which was placed in the irradiation position shown. A fall in the absorbed dose of a factor of 200 over a distance of approximately 3 cm, was measured in this way.

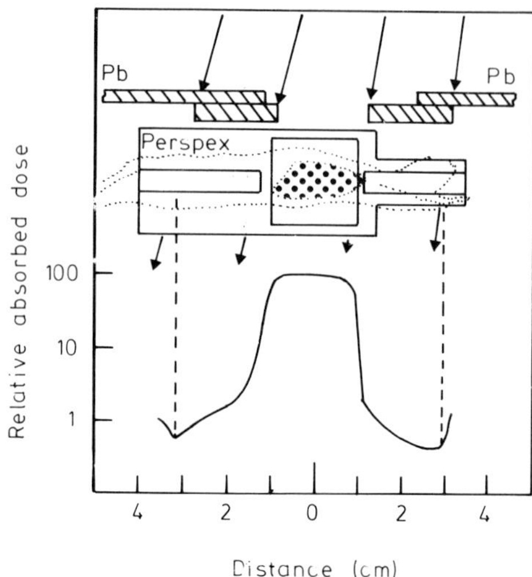

Figure 4.4 Absorbed dose distribution along the central axis of the body of a mouse irradiated in the thoracic region. The lung is shown shaded. (Scarpa *et al* 1979, reprinted with the permission of the *British Journal of Radiology*.)

4.3 Radiotherapy Absorbed Dose Measurement

4.3.1 Simple-geometry phantoms

In radiotherapy the specification of the complete absorbed dose distribution within the radiation beam in a phantom is a prerequisite for calculating the prescribed absorbed dose to the target volume in the patient. The commonest method of specifying this is to use published depth dose data and an isodose chart. An example of such a chart is illustrated in figure 4.5. This chart refers to a section containing the beam axis parallel to one side of a ^{60}Co 10 × 10 cm therapy beam for a fixed source to skin distance (SSD) of 80 cm. The lines mapped on the chart link points of equal absorbed dose expressed as a percentage of the peak absorbed dose. In the case of ^{60}Co the peak absorbed dose occurs at the optimum 'build-up depth' in water (5 mm). For x-ray beams with generating potentials of less than 400 kV the depth doses are conventionally expressed as a percentage of the surface absorbed dose. Similar charts are also used for fixed source–axis distance (SAD) beams, although in these the isodose values are expressed as a percentage of the absorbed dose at the target deep within the phantom. SAD isodose charts are used when the target volume is located on the axis of rotation of the teletherapy machine.

The selection of an appropriate isodose chart can be difficult since the absorbed dose distribution in the phantom depends on the beam dimensions, SSD or SAD, the radiation quality, the source size, the geometry of the beam, and the positioning of the beam collimators. The International Commission on Radiological Units (ICRU 1976) therefore recommends the use of isodose charts which are exactly specified for the particular equipment being used. This criterion can be established, as ICRU recommend, by a series of single measurements using an ion chamber or TL dosemeters.

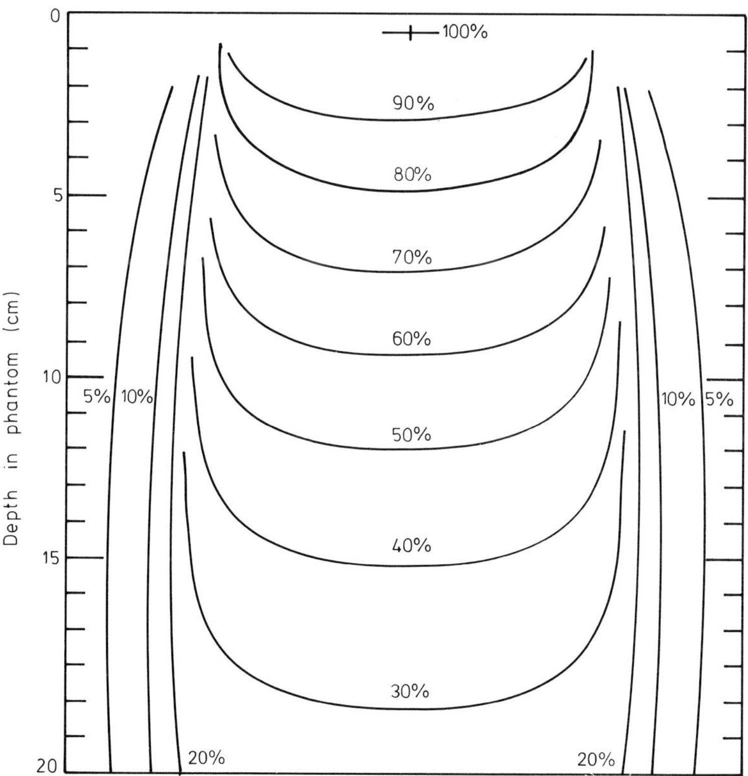

Figure 4.5 Radiotherapy isodose chart for a ^{60}Co teletherapy unit. Field size 10 × 10 cm, source to skin distance 80 cm.

In their simplest form the measurements consist of

(1) measuring the depth dose distribution along the central axis of the beam in a water or water-equivalent phantom;

(2) choosing one particular phantom depth which, in the case of 150 kV to 10 MeV ICRU recommends as 5 cm, and measuring the radiation absorbed dose profile across the beam at this depth.

After normalisation of the depth dose measurements at 5 cm depth, the published depth dose data which are to be used can be compared with them and corrected accordingly. Similarly, the measured beam profile can be compared with that obtained from the published isodose chart. These measurements do not of course provide information about the absolute value of absorbed dose rate delivered to a position within the phantom, but only relative values of absorbed dose distribution. The absorbed dose rate can only be obtained by direct comparison with measurements obtained using a calibrated ionisation chamber.

Since 1968 the International Atomic Energy Agency (IAEA) and the World Health Organisation (WHO) have been running a programme of intercomparison of ^{60}Co teletherapy units in the various radiotherapy centres throughout the world. This resulted from investigations carried out in 1965 which revealed that there was no suitably calibrated radiation measuring instrument in use in about 30% of the radiotherapy centres investigated (Eisenlohr and Jayaraman 1977). A simple test procedure based on TL dosimetry was initiated to assess the accuracy of delivered absorbed doses in the centres.

The TL measurement technique used in this study illustrates:

(1) the practical use of TL dosemeters for radiotherapy depth dose measurements in a simple phantom; and

(2) methods to eliminate effects of fading and other variable environmental factors.

The procedure used is illustrated in figure 4.6.

Participant radiotherapy centres are sent four sets of LiF:Mg:Ti powder dosemeters contained in PTFE capsules. They are requested to irradiate one test set (A) with an absorbed dose of 2 Gy in water at 5 cm depth on the central axis of a ^{60}Co 10×10 cm therapy beam with an 80 cm SSD. Another test set (B) is to be irradiated under similar conditions for 2 min. A control set (C) which has been given a known absorbed dose by IAEA and a control set (D) which is unexposed, accompany sets A and B at all times except during irradiation. Sets C and D provide information about any environmental or spurious effects, such as thermal

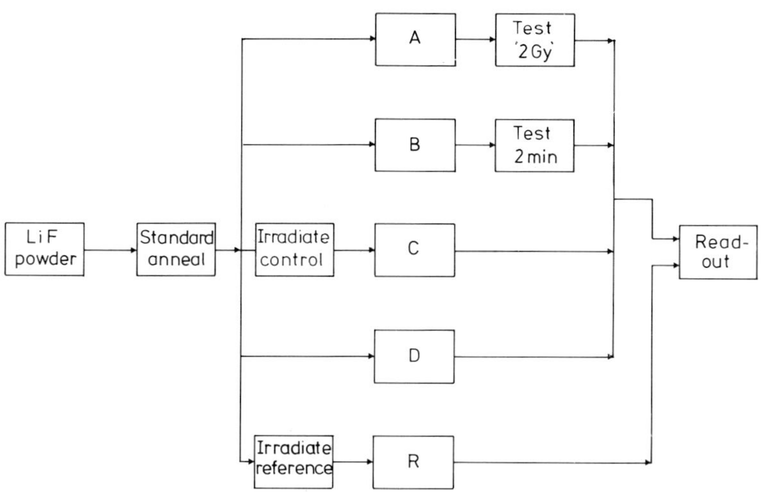

Figure 4.6 TLD method employed by IAEA and WHO for the intercomparison of delivered absorbed doses from ^{60}Co teletherapy units in various radiotherapy centres throughout the world. A and B, test measurement dosemeters; C and D, irradiated and unirradiated control dosemeters; R, calibration reference dosemeters (Eisenlohr and Jayaraman 1977).

fading, unintentional irradiation, etc, which might adversely affect the test dosemeters. In addition reference sets (R) are irradiated by IAEA in a standardised 10 × 10 cm ^{60}Co beam (SSD 80 cm) at a depth of 5 cm in water. The absorbed dose rate expressed in Gy min^{-1} in water is obtained from a measurement of the exposure rate in free air using a calibrated ionisation chamber. All dosemeters are then read out together, eliminating possible calibration errors due to fading, and standardising the readout procedure for all dosemeters. The results of these TL measurements, together with an analysis of the absorbed dose calculation procedures adopted by the various centres, have proved very useful in identifying errors in the assessment of treatment absorbed dose and in improving the overall accuracy of dosimetry within certain centres.

The versatility in dosemeter design which can be achieved by the user of TL dosemeters is exemplified by an interesting technique developed to measure depth dose and isodose curves in a polystyrene phantom irradiated with a ^{90}Sr/^{90}Y applicator source (Rudén and Bengtsson 1974). Fine spatial resolution was obtained using LiF:PTFE disc-shaped dosemeters 1 mm in diameter and 0.1 mm thick, which were prepared by punching out from the normal-size LiF:PTFE discs. Linear arrays of dosemeters were made by inserting individual discs in 0.1 mm deep grooves on the surface of 1 mm thick polystyrene wafers. The shuttered applicator source was inserted in a hole in the polystyrene, and the dosemeters were then placed in their measurement positions (as shown in figure 4.7), and irradiated. Distances between source and dosemeters could be varied by the insertion of a series of 1 mm thick polystyrene wafers. Dosemeter response factors were measured within a polystyrene phantom using ^{60}Co photons. The absorbed dose (Gy in polystyrene) was calculated by converting the measured exposure rate in free air and applying the appropriate phantom–air ratio, and Gy kg C^{-1} factors. (The TL response of LiF:PTFE to ^{60}Co and $^{90}Sr/^{90}Y$ radiations is the same to within a few per cent.)

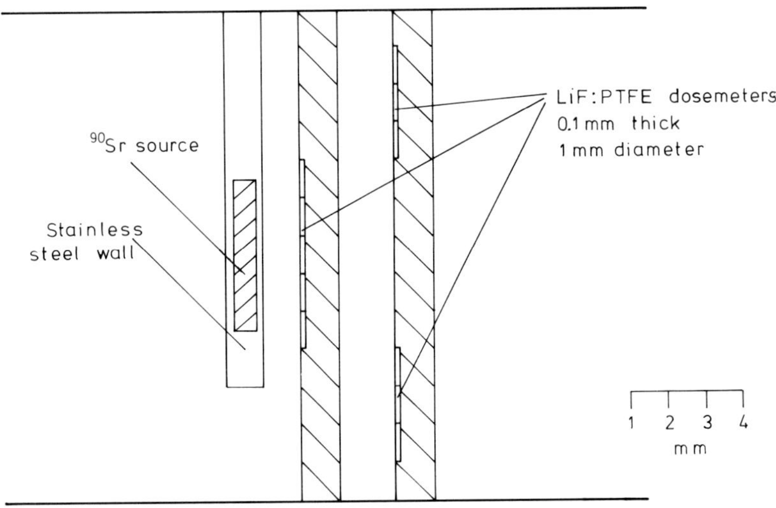

Figure 4.7 Source–TL dosemeter configuration for the measurement of depth dose in a polystyrene phantom. Only one array of dosemeters was used for each measurement (Rudén and Bengtsson 1974).

Isodose curves obtained by Rudén (figure 4.8) illustrate the fine degree of spatial resolution afforded by TL dosemeters for phantom measurements and the excellent agreement between measured and calculated absorbed dose rates.

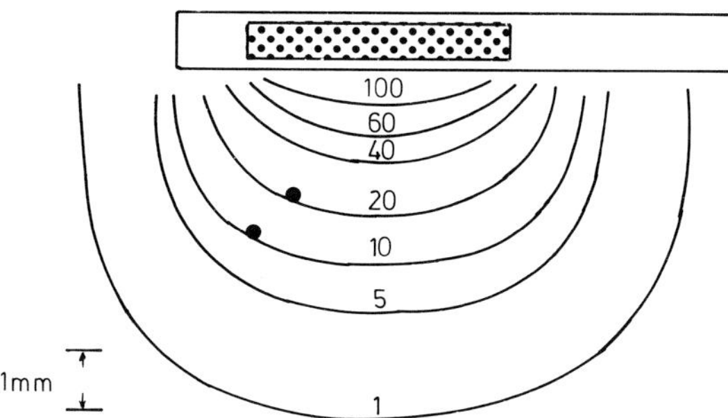

Figure 4.8 Isodose curves obtained using TLDs (1 mm diameter, 0.1 mm thick LiF:PTFE disc dosemeters) in a polystyrene phantom in the measurement of absorbed dose distribution around a ^{90}Sr applicator. Absorbed dose rates are shown as Gy min^{-1}. The dots represent theoretical values (Rudén and Bengtsson 1974).

4.3.2 *In vivo* measurement

While the measurement of complete absorbed dose distribution in a phantom is essential in planning the treatment of a patient, the ultimate check on the absorbed dose delivered to the patient can only be made by *in vivo* absorbed dose measurements. TL dosemeters have proved to be particularly useful for this purpose.

The relevance of *in vivo* dosimetry is illustrated by the flow chart shown in figure 4.9. This flow chart is a simplified form of that used by

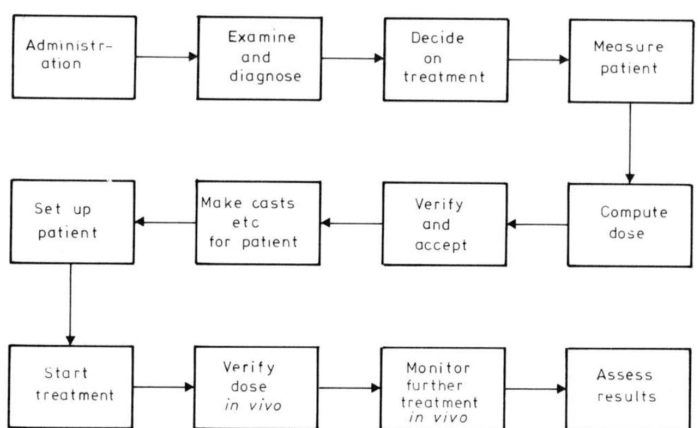

Figure 4.9 Flow chart illustrating the relevance of *in vivo* dosimetry in radiotherapy planning and treatment.

ICRU (1976) to illustrate a systems approach to radiotherapy. *In vivo* measurements verify that the absorbed dose prescribed by the clinician, and calculated and set up by the physicist and the radiographer, has been delivered. Further it may be used to monitor any change in field uniformity caused by changes in the many treatment parameters. Such is the complexity of many radiotherapeutic procedures that significant uncertainties can arise.

In vivo measurements can be divided into four classes.

(i) *Entrance absorbed dose measurements.* These are used mainly to check the machine output, the absorbed dose distribution profile across the patient (particularly in the penumbra of shielding) and the positioning of shielding in relation to the position of the patient. If the measured values are at variance with those prescribed and calculated, the cause can be investigated and appropriate corrective action taken. The spatial resolution afforded by TL dosemeters is particularly useful in these measurements.

(ii) *Exit absorbed dose measurements.* These are used mainly to check the absorbed dose delivered to points deep within the body. The measurements should agree with calculations for exit absorbed doses. For such measurements the TL dosemeters should be provided with sufficient backscatter material, and again good spatial resolution is useful.

(iii) *Intra-cavitary absorbed dose measurements.* The absorbed dose within a body cavity (e.g. the mouth, nasopharynx, oesphagus, vagina, rectum, etc) can be measured using TL dosemeters sealed inside a catheter, as shown in figure 3.9. The position of the dosemeters may be checked using radio-opaque markers and exposing an x-ray film. It should be noted that the increase in scattered radiation resulting from the presence of high atomic number radio-opaque markers can cause uncertainties of a few per cent in the absorbed dose to the TL dosemeter, although this can be allowed for.

(iv) *Individual spared organ absorbed dose measurements.* The absorbed dose to spared (shielded) organs can be measured but often no build-up can be used as this would in itself result in an increased absorbed dose to the organ.

TL dosemeters are routinely used for a number of different types of *in vivo* radiotherapy measurements in centres throughout the world.

The following examples illustrate some of the principles and detail some of the techniques used.

4.3.3 Measurement of absorbed dose during 'mantle' therapy for Hodgkin's disease

The radiotherapy treatment of Hodgkin's disease involves the irradiation of a large area of the body. The treatment field is designed to deliver a therapeutic or prophylactic absorbed dose to the axillary, cervical and

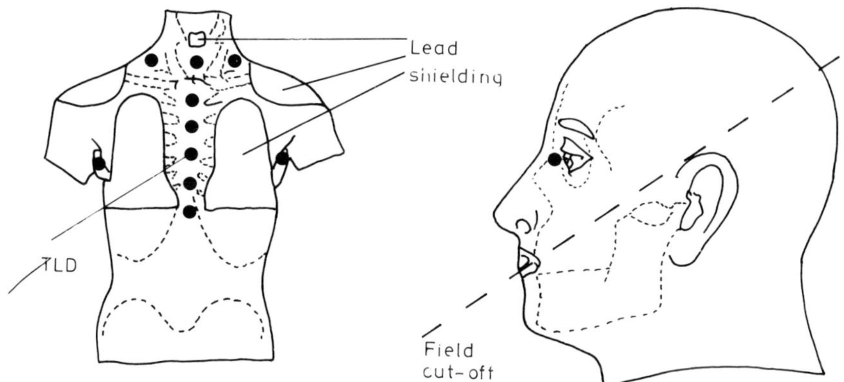

Figure 4.10 Mantle treatment for Hodgkin's disease. Shielding is shown as a projection onto the body surface.

mediastinal lymph nodes. Many different treatment fields have been used but one commonly used configuration consists of anterior and posterior parallel and opposed fields, as illustrated in figure 4.10. There is also the need to shield presumed healthy organs such as lips, eyes, lungs, kidneys, bone joints, etc, and, for posterior irradiation, the spinal cord. To avoid excessive exposure of the skin (skin sparing) the treatment is carried out using either ^{60}Co or megavoltage x-ray photons and the shielding is positioned some distance (typically 20–50 cm) above the entrance surface of the body. This is achieved using 'individually tailored' moulds of polystyrene, one anterior and one posterior, with appropriately cut out channels containing lead shot shielding. Alternatively, appropriately shaped lead absorbers, placed on a Perspex plate and positioned above the body, have been used. The technique is termed the 'gothic arch' or 'mantle' technique. A typical course of treatment involves a total prescribed absorbed dose of between 30 and 40 Gy delivered in 20 fractions over a four-week period.

TL dosemeters, usually LiF extruded ribbons or rods, or PTFE-based discs or micro-rods sealed in thin protective polythene sachets, are attached to the body under moulded blocks of wax or in small Perspex containers. The dimensions of the wax or Perspex are chosen to provide build-up appropriate to the photon energy of the beam, and to ensure electronic equilibrium. With the use of high-energy photon radiation (e.g. 42 MV x-rays) Rudén (1976) recommends the use of a maximum build-up of 15 mm of Perspex and an experimentally derived factor to correct the apparent absorbed dose. Exit absorbed dose measurements can be made in a similar manner or by attaching the dosemeters directly onto the patient's immobilisation cast.

The results of such a series of measurements of entrance and exit absorbed doses for mantle fields are shown in table 4.3.

Table 4.3 Comparison of *in vivo* TL measured and calculated (in parentheses) entrance and exit absorbed doses for 'mantle' therapy fields (Suntharalingam and Mansfield 1971). M, mediastinum; A, axilla; L, lung.

TL dosemeter (shielding)	Area	Absorbed dose (Gy)	
		Entrance	Exit
LiF: PTFE disc (Pb shot)	M	3.14 (2.90)	0.99 (0.93)
	A	2.55 (2.70)	1.30 (1.10)
	L	0.41 (0.32)	0.40 (0.40)
LiF ribbons (Pb blocks)	M	2.92 (2.83)	0.84 (0.93)
	A	2.48 (2.52)	0.89 (0.93)
	L	0.25 (0.30)	0.32 (0.35)

4.3.4 Measurement of relative skin exposure distribution

Radiations, with limited penetration in tissue, are routinely used for partial or whole body radiotherapy treatment of certain skin conditions. Such treatments include the use of Grenz rays (x-rays 5–20 kV, 0.01–0.035 mm Al HVL), superficial x-rays (60–120 kV, 0.5–2 mm Al HVL), beta plaque sources (e.g. $^{90}Sr/^{90}Y$), and electron beams (DHEW 1977). The use of such radiations allows relatively large deposition of energy in the outer few millimetres of skin with minimum penetration of underlying organs.

Thermoluminescence dosemeters provide a method of measuring the spatial distribution of skin exposure over the treatment area and the exposure to shielded organs such as the eyes, thyroid, gonads and breasts. Dosemeters such as LiF powder, extruded ribbons and PTFE discs or $Li_2B_4O_7$:Mn powder or extruded ribbons, sealed in thin protective polythene sachets, may be attached to the skin, as illustrated in figure 4.11. The positions of thermoluminescence dosemeters in relation to six fields (three anterior, three posterior) used for the treatment of mycosis fungoides are shown in figure 4.12; shielded organ shielding is not shown. In this case, superficial x-rays (100 kV, 1.0 mm Al HVL) are used and the total two-week treatment exposure of 100 R (2.58×10^{-2} C kg^{-1}) was delivered in ten equal fractions. The boundaries of the treatment fields were marked on and around the patient by linear

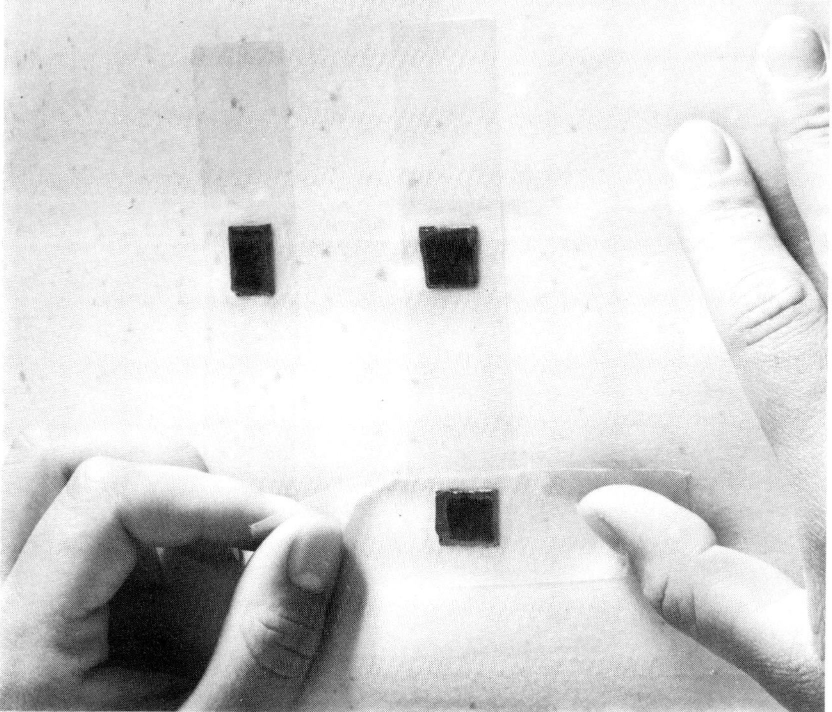

Figure 4.11 Black polythene sachets containing $Li_2B_4O_7$:Mn powder being attached to the skin of a patient. (Photograph courtesy of the National Radiological Protection Board, Harwell.)

extension of the angled sides of the applicator cone. This procedure can lead to uncertainties in field cut-off positions and possible overlapping of treatment fields. The exposure of areas such as the thighs is particularly difficult to predict.

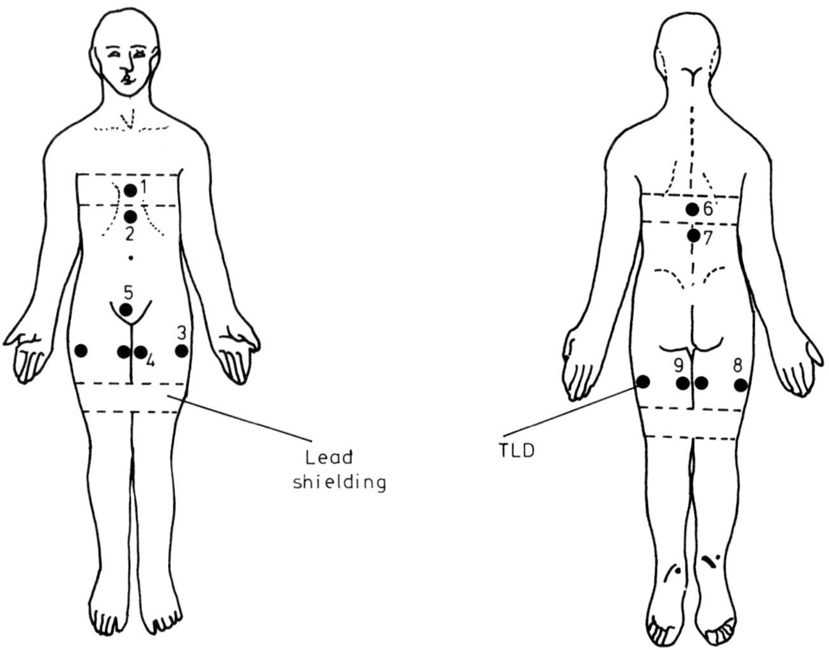

Figure 4.12 Fields and TLD arrangement for mycosis fungoides irradiation.

Results of *in vivo* measurements taken during the first fraction of a series of treatment exposures are presented in table 4.4. They indicate the significant uncertainties which can arise and, in particular, the requirement for supplementary exposure of the thighs. The series of *in vivo* measurements during all treatment fractions enabled the cumulative exposure to all areas of skin to be determined. At lower photon energies the effects of dosemeter orientation and self-shielding discussed in § 7.4 will become increasingly important.

4.3.5 Intra-cavitary absorbed dose measurement

The small size and shape of the extruded and PTFE-based micro-rod TL dosemeters have enabled *in vivo* measurements of absorbed dose inside body cavities which hitherto were often difficult and sometimes impossible. As illustrated in figure 3.12, these dosemeters can be easily inserted and sealed in catheters.

A very good example of such applications is the *in vivo* measurement of absorbed dose distribution in the pelvis during intra-cavitary ^{226}Ra and external beam therapy for carcinoma of the uterine cervix (Johansson *et al* 1969, Joelsson and Backström 1970, Joelsson *et al* 1972).

Table 4.4 *In vivo* measurements of relative distribution of skin exposure for a single treatment fractional exposure of 10 R (2.58×10^{-3} C kg^{-1}).

Field	Dosemeter position†	Percentage of prescribed exposure
	1‡	4
Anterior	2	106
SSD 113 cm	3	22
	4	5
	5	82
Posterior	6‡	2
SSD 110 cm	7	85
	8	35
	9	47
SSD 110 cm	Central axis calibration	100

† Dosemeter positions relate to figure 4.12.
‡ Under lead shielding.

The dosemeters are introduced into the pelvis via the external femoral veins. In one technique (Johansson *et al* 1969) sterile PTFE catheters are first inserted into the veins, and the process is monitored using x-ray fluoroscopy and television. Inner tubes containing gold radio-opaque markers are introduced into the catheter to assess accurately the intended positions of the TL dosemeters; the radium is applied and its position in relation to the dosemeter markers can be assessed; the marker catheter tubes are then removed and replaced by two others each containing 15 micro-rod dosemeters spaced some 16 mm apart. Using this technique of inner and outer catheters there is no need to sterilise the inner dosemeter catheters.

Table 4.5 Intra-cavitary pelvic dosimetry during treatment for carcinoma of the uterine cervix. Résumé of measured absorbed doses obtained from the TL dosemeters introduced into the left and right femoral veins. Calculated values are in parentheses (Johansson *et al* 1969).

Absorbed dose ($Gy\,kg^{-1}\,h^{-1}$)	
Left	Right
1132	648
1772 (1743)	925
1582 (1471)	902 (914)
748 (764)	713 (657)
428 (421)	455 (414)
248 (243)	260 (236)
201	198

The results of two series of such measurements using this technique (table 4.5) verify the computed treatment absorbed dose and provide useful information about the symmetry of the absorbed dose distribution which correlates with the relative positions of irradiators and TL dosemeters.

4.4 Diagnostic Radiology Absorbed Dose Measurement

In the diagnostic range of photon energies LiF phosphor can over-respond by as much as 40% compared with tissue. However, $Li_2B_4O_7$:Mn is an extremely good match for tissue over this range of photon energies and, by adjustment of the fractional amount of manganese present, the response can be 'trimmed' to match more closely that of air, water or tissue. For example, Jayachandran (1970) suggested 0.34% wt/wt for air equivalence and Christensen (1967) 0.45% wt/wt. Langmead and Wall (1976), using $Li_2B_4O_7$:Mn powder containing 0.15% wt/wt of manganese, found that they could measure the absorbed dose in tissue, from x-rays of unknown quality, with a predicted error associated with photon energy of $\ngtr \pm 5\%$ and an overall uncertainty of $\ngtr \pm 15\%$. This is an important characteristic of $Li_2B_4O_7$:Mn phosphor because, while the effective energy of the primary beam can be assessed (at least in air) by half-value layer measurements, the quality of the lower-energy scattered radiation and the magnitude of its contribution to the absorbed dose are difficult to assess. $Li_2B_4O_7$:Mn is therefore particularly suited for such measurements. While sensitive solid forms of $Li_2B_4O_7$:Mn, such as extruded ribbons, are commercially available, they are relatively expensive, and this tends to exclude them from large-scale measurement programmes. Loose powder is probably most suitable for this application at present. Measurements of absorbed dose in the diagnostic range of photon energies can of course be carried out using other phosphors but greater care is required to correct for the effect of photon energy.

4.4.1 Anthropomorphic phantom measurements

A number of absorbed dose measurements in simple-geometry phantoms has been described in § 4.3.1; for example, the measurement of the beam profile of a therapy machine.

In contrast with these simple homogeneous phantoms a most useful phantom for absorbed dose measurements in radiotherapy and diagnostic radiology is one which is designed, as far as is practicable, to simulate the structure of the human body (anthropomorphic). The torso, and head and neck of such a phantom are shown in figure 4.13. Proportionally it is equivalent to an 'average man' 1.75 m tall and weighing 73.5 kg. It is made from tissue-equivalent synthetic rubber and contains a complete human skeleton, lung-equivalent material, and airways corresponding with the maxillary sinuses, nasopharynx, trachea, etc. It is sliced into 25 mm thick transverse sections each containing a matrix of 5 mm diameter holes spaced 3 cm apart. Each hole can accommodate a dosemeter holder/capsule or a solid plug of tissue-equivalent material. The complete phantom contains over 3000 holes and additional ones can be drilled if required. Suitable Perspex or polythene capsules can each contain approximately 35 mg of powdered TL phosphor.

Anthropomorphic phantoms are extremely useful for absorbed dose measurements in diagnostic radiology where their human form not only allows precise and realistic positioning of 'the patient' in the beam, but also the positioning of dosemeters to measure the absorbed dose to specific organs, including the gonads.

The main useful features of (1) film, i.e. its spatial continuity and permanent recording ability, and (2) LiF or $Li_2B_4O_7$:Mn TL phosphors, i.e. their approximate tissue-equivalence over the diagnostic photon energy range, can be combined. Vacirca *et al* (1972) developed such a technique for the measurement of body absorbed doses from diagnostic x-rays. This enables one to obtain a continuous absorbed dose distribution across each transverse section and throughout the entire volume

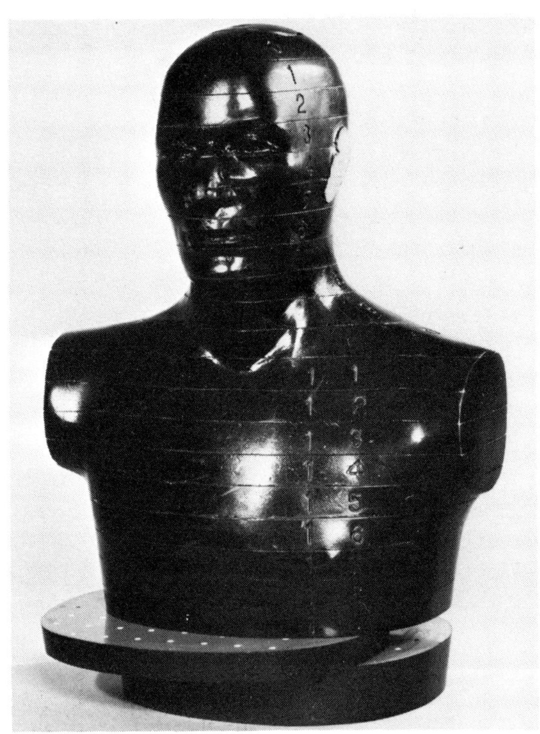

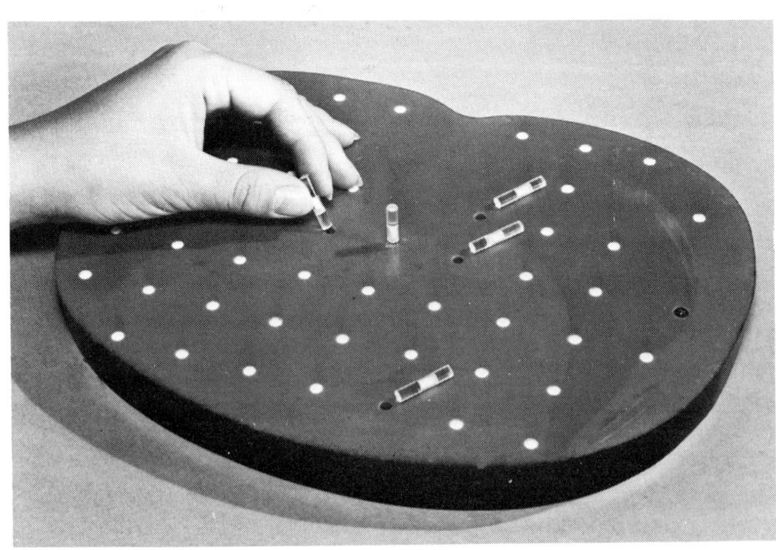

of a phantom. Vacirca's method consisted of placing x-ray film packs and individually calibrated (relatively) TLD 100 LiF extruded-rod dosemeters together, between the transverse sections of a phantom. The phantom was then irradiated under realistic simulated radiographic examination conditions as illustrated in figure 4.14(*a*). A range of different exposure times must be used to ensure adequate film darkening and TL signal whether the dosemeters lie in or out of the useful beam. The *continuous* absorbed dose distribution across any section of a phantom can be obtained in this way by measuring the optical density of the developed film at any chosen point. Calibration of the film is achieved by comparing the TL reading with a measurement of film optical density at each TL dosemeter site, and calibration curves relating to film optical density and TL signal can then be drawn, as illustrated in figure 4.14(*b*).

Isodose curves, expressed as a percentage of the axial entrance absorbed dose, are similar to, but provide more spatial detail than those constructed from the fewer point TL measurements. The absorbed dose at any chosen site within the phantom can be determined from the percentage depth dose datum for the site and the axial exposure rate at the entrance surface of the phantom. Results of measurements obtained at different sites within the phantom using this method are shown in table 4.6.

Table 4.6 Relative absorbed dose (expressed as a percentage of axial entrance absorbed dose) to sites within a phantom measured using a combination of films and TLDs. The beam geometry is illustrated in figure 4.14(*a*) (Vacirca *et al* 1972).

Site	Beam area	Beam quality	
		100 kVp, 3.9 mm Al HVL	80 kVp, 2.8 mm Al HVL
Skin	Useful beam	100	100
Exit		4.9	3.6
Sternum		11.0	8.1
Ovary	Scattered beam	0.077	0.044
Scrotum		0.035	0.025

Figure 4.13 Anthropomorphic phantom made from tissue-equivalent synthetic rubber. Lower photograph shows dosemeter capsules, each containing 30 mg of phosphor being inserted into a transverse section of the phantom. (Photographs courtesy of the National Radiological Protection Board, Harwell.)

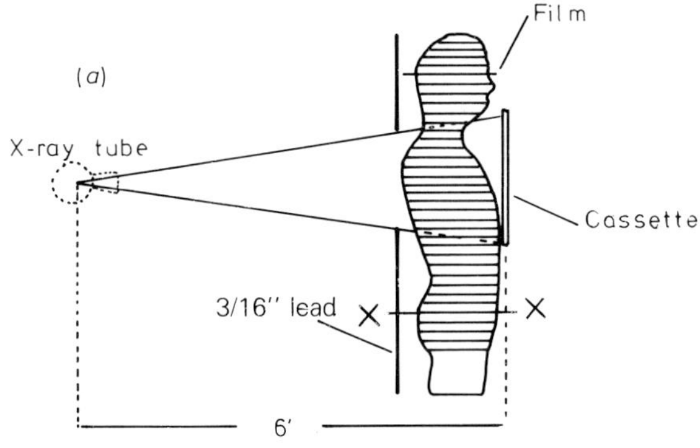

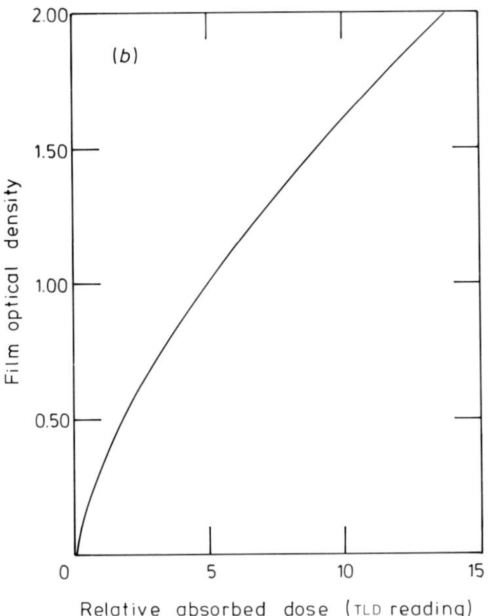

Figure 4.14 Film–TLD method for the measurement of absorbed dose from diagnostic x-ray exposure. (*a*) Beam geometry for simulated chest x-ray. Films and TLDs are placed between the shaded transverse sections. (*b*) Calibration curve of film optical density against TL emission for film and TLDs in the plane ×–× shown in (*a*) situated 7 inches below the lower cut-off limit of the useful beam (Vacirca *et al* 1972).

Realistic phantoms combined with TL dosimetry have also proved useful in the assessment of 'patient' absorbed dose imparted by computerised axial tomography (CAT) (e.g. Horsley and Peters 1976, Wall *et al* 1979b). The recent measurements by Wall *et al* were performed using $Li_2B_4O_7$:Mn powder dosemeters contained in plastic containers and inserted in the phantom slices, as illustrated in figure 4.13. In these examinations, regions of dosimetric interest include not only the section of the patient (phantom) undergoing radiological examination at a particular instant in time, but also the adjacent sections which are irradiated as a result of the divergence and scatter of the primary beam. Wall *et al* (1979b) also included measurements of absorbed doses to the lens of the eye, thyroid, gonads and skin. For these they used $Li_2B_4O_7$:Mn powder in polythene sachets.

4.4.2 *In vivo* measurements

While anthropomorphic phantom measurements are extremely useful in the assessment of absorbed dose in diagnostic radiology the measurements are not performed under realistic conditions. The information gained from these measurements takes no account of the skill and experience of the radiographer, the quality and suitability of equipment, nor differences in the shape and size of patients. *In vivo* measurements on patients undergoing routine radiological examinations in hospitals provide a much more realistic assessment of absorbed dose under practical conditions.

$Li_2B_4O_7$:Mn dosemeters are especially useful for these measurements. As we have seen, they are tissue-equivalent and radio-transparent even

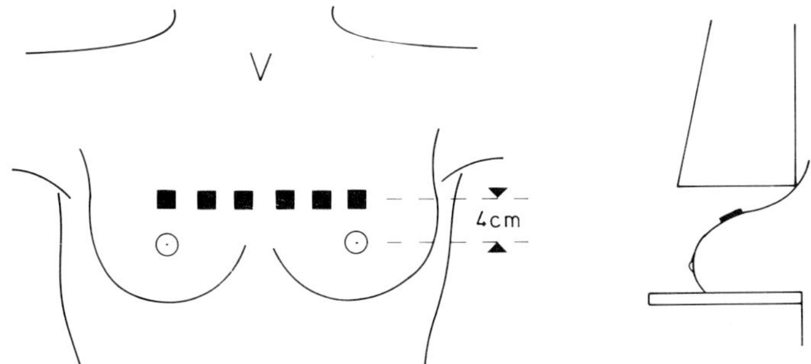

Figure 4.15 Arrangement of $Li_2B_4O_7$:Mn dosemeters (shown in figure 4.12) for mammography measurements. (Langmead *et al* 1976, reprinted with the permission of the *British Journal of Radiology*.)

to very low-energy diagnostic x-rays, such as those used in mammography and in consequence they do not interfere with the diagnostic quality of the image and cause little inconvenience to patient, radiographer and radiologist.

Langmead *et al* (1976) used $Li_2B_4O_7$:Mn powder dosemeters (the same as those shown in figure 4.11) for the measurement of absorbed doses to patients undergoing various forms of radiological examination, including cardiac catheterisation, barium enemas, intravenous pyelography and mammography. Maximum skin and gonad absorbed doses were measured. The measurement positions of the 1×1 cm dosemeters for mammography are shown in figure 4.15.

This series of measurements constituted a pilot survey of absorbed doses to patients and this work has recently been extended to include other radiological techniques.

5 Other Applications of TLD

5.1 Individual and Environmental Monitoring

5.1.1 Individual monitoring (personal dosimetry)

The International Commission on Radiological Protection (ICRP) divide conditions of work involving ionising radiations into two classes (ICRP 1977):

Class A where the annual exposures might exceed three-tenths of the dose-equivalent† limits, and

Class B where it is most unlikely that the annual exposures will exceed three-tenths of the dose-equivalent limits.

The dose-equivalent limits are those values of dose-equivalent set to prevent non-stochastic effects‡ and to limit stochastic effects to an acceptable level.

The ICRP recommend that workers performing work of class A category should be subject to individual dose monitoring for external radiation, and for internal contamination where appropriate. For workers whose work falls within class B individual monitoring is not necessary, but rather the conditions in the working environment are assessed.

In general, monitoring of absorbed doses to individuals may serve to

(1) ensure that dose-equivalent limits are not exceeded;
(2) limit the exposure of individual workers;
(3) assist in retrospective analysis in the event of an accidental over-exposure; and

† Dose-equivalent (H) is used only for radiation protection purposes and is defined as the product $H = QND$, where Q is the quality factor appropriate to the radiation, D is the absorbed dose and N is the product of all other modifying factors; for external radiation $N = 1$. The units of dose-equivalent are sieverts (1 Sv = 100 rem).

‡ Stochastic effects are those whose probability of occurrence is a function of absorbed dose, without threshold. Non-stochastic effects are those whose severity is a function of absorbed dose and for which a threshold may occur.

(4) provide supplementary information about work practices and dose trends.

In the past, individual doses have almost always been monitored by film badge dosemeters but TLDs are now becoming widely used. This has been due principally to the availability of automated readout systems which can be linked to computer dose record-keeping systems. Large, centralised automated TLD systems are being developed in a number of countries (Dennis *et al* 1974, Duftschmid 1980, Grogan *et al* 1980).

5.1.1.1 Requirements of TLDs for individual dose monitoring. The CEC (1975a, b) have set out general recommended requirements for TLDs for use in individual monitoring. In addition a standard specifying performance criteria for personal and environmental TLDs, together with recommended procedures for testing compliance with the performance criteria, is being prepared by the International Standards Organisation. As this document is presently available only as a draft and the information contained therein is highly specific, it will not be considered further in this text.

Three types of dosemeter are recommended by the CEC: (i) non-discriminating dosemeters; (ii) discriminating dosemeters; and (iii) extremity dosemeters.

(i) *Non-discriminating dosemeters.* This is the simplest form of dosemeter providing a measure of the absorbed dose only, with no information about radiation quality or direction. If it is established that only *either* penetrating *or* non-penetrating radiation is present it is, in principle, possible to use a dosemeter containing a single tissue-equivalent TL element.

The division between penetrating and non-penetrating radiation is somewhat vague. Radiation which, by virtue of its rapid attenuation and absorption by the surface tissues of the body, produces negligible effect in deeper tissues, is defined as non-penetrating. The skin is then the only organ at risk. Penetrating radiation, on the other hand, irradiates deeper-lying tissues and organs—in particular, the gonads and bone marrow. With some radiations, such as certain β-rays, the distinction is unclear. They can irradiate not only the skin but also some deeper tissues such as the lens of the eye, although they have almost always negligible effect on the gonads and bone marrow. It is therefore convenient to define measurement depths specific to particular organs. These are usually specified as depth in tissue (mm) or equivalent mass per unit frontal area (mg cm^{-2}).

The CEC (1975b) recommend the use of a two-element (tissue-equivalent) dosemeter to provide only basic absorbed dose information. One element should measure the skin absorbed dose at the depth of the basal layer cells, i.e. between 5 and 10 mg cm^{-2}. Ideally, what is required is a 5 mg cm^{-2} tissue-equivalent TL element underneath 5 mg cm^{-2} of tissue-equivalent absorber. The other element should measure the body absorbed dose at a depth of between 400 and 1000 mg cm^{-2} (approximate range of depths of male gonads). The use of this element will result in a safe overestimate of female gonad absorbed dose (approximate depth 7000 mg cm^{-2}) and bone marrow absorbed dose (approximate depth 2000 mg cm^{-2}). While it is true that the absorbed dose to deeper organs and tissues will be overestimated, this is not necessarily the case for the eye.

In practical dosemeters the dosemeter elements are almost always the result of a compromise between the ideal requirement and the realities of manufacture and use. Thin, robust skin dosemeters are particularly difficult to produce.

The measurement of skin absorbed dose from exposure to low-energy β-radiation is particularly difficult using TLDs. The use of dosemeters thicker than the recommended 5–10 mg cm^{-2} results in an underestimate of absorbed dose. Charles (1977) produced ‘ultra-thin bonded disc dosemeters’ with a thickness of approximately 6 mg cm^{-2}, by thermally bonding 25 μm thick phosphor-loaded PTFE discs to 0.2 or 0.5 mm thick pure PTFE discs. The phosphors used were LiF : Mg,Ti and $CaSO_4$: Dy. However high and variable backgrounds, mainly associated with light-induced TL in the PTFE, limited the thresholds of absorbed dose to 0.8 and 2 mGy for the $CaSO_4$: Dy and LiF : Mg,Ti, respectively. In this respect they do not conform to the CEC recommended performance.

Other approaches to the problem of low-energy β absorbed dose measurement have included

(1) dosemeters loaded with graphite to absorb the TL signal originating from deep inside the dosemeter—reading out only TL from the ‘surface layer’ (Koczynski *et al* 1974), and

(2) dosemeters formed by diffusing boron into LiF to a depth equivalent to 1–2 mg cm^{-2} (Christensen and Mäjborn 1980). That part of the dosemeter containing the boron has a characteristic glow peak at 340 °C.

(3) dosemeters produced by sticking a uniform thin layer of phosphor powder onto a substrate.

Benko *et al* (1977) adopted the last approach producing a 10 mg cm^{-2} thick $CaSO_4$: Dy dosemeter in the form of powder (30 μm grains) deposited on an aluminium disc. The minimum detectable absorbed dose was 50 μGy and the dependence of TL sensitivity on β energy was less than ± 15% over the range of maximum β energies 200–2000 keV. The β energy dependence of this and a number of other dosemeters is shown in figure 5.1.

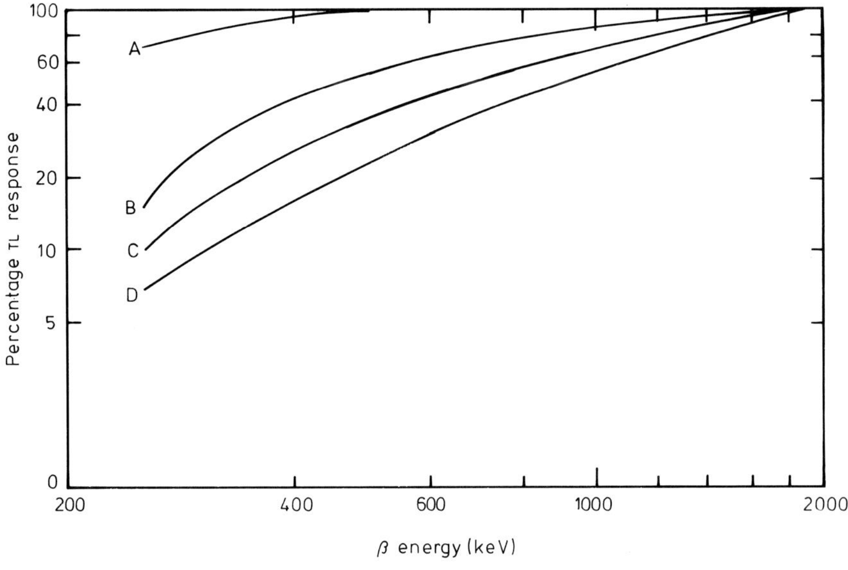

Figure 5.1 TL response of some phosphors to β-radiation. A, $CaSO_4$:Dy on an aluminium disc; B, LiF (0.6 mm); C, LiF (0.76 mm chip); D, CaF_2Mn (1 mm rod). (Data from Benko *et al* 1977.)

(ii) *Discriminating dosemeters.* In addition to the basic absorbed dose information, the discriminating dosemeter should provide further information about the quality and direction of the radiation. Such information can be useful in the event of accidental over-exposure, where it is important to identify the source of the radiation. Hitherto such qualitative information has been available using film detectors in combination with a badge containing a range of metal and plastic filters. Information about radiation quality in the energy range 40–200 keV can be obtained in principle using a TLD comprising (*a*) a system of metal filters and at least two TL elements (for example, as illustrated by the relative photon energy response of TLD 100 extruded-ribbon elements shielded by a range of metal filters as shown in figure 5.2), or (*b*) at least two elements, one tissue-equivalent and the other highly energy-dependent.

The first method, although well established in film dosimetry, has the disadvantage of introducing a marked angular dependence. To provide the qualitative information equivalent of the film badge would require many TL elements and would be expensive.

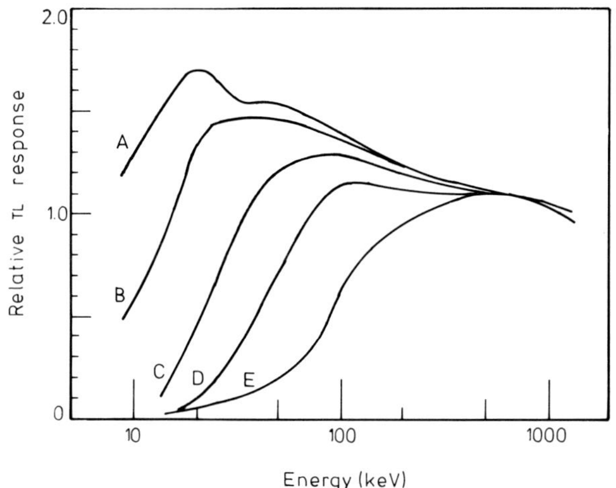

Figure 5.2 The relative energy response of LiF TLD 100 ribbons (relative to ^{60}Co) covered by different filters. A, open window; B, 3.0 mm plastic; C, 2.0 mm aluminium; D, 0.3 mm copper; E, 1.0 mm tin (Julius 1976).

(iii) *Extremity dosemeters.* Where it is believed that the body dosemeter cannot give a true indication of the absorbed dose to an extremity such as the hands, that is, where during the working period the absorbed dose to the extremity is likely to be significantly higher than that to the body, an extremity dosemeter can also be worn. An example is shown in figure 5.3. The TLD phosphor—in this case LiF:Mg:Ti powder—is contained in a sachet or finger stall. The particular configuration chosen will depend on the manual task being performed and the resultant estimate of absorbed dose rate distribution across the hands. There are at present no internationally agreed recommended performance criteria specific to extremity dosemeters. Up to 65 000 extremity dosemeters are used annually in the UK.

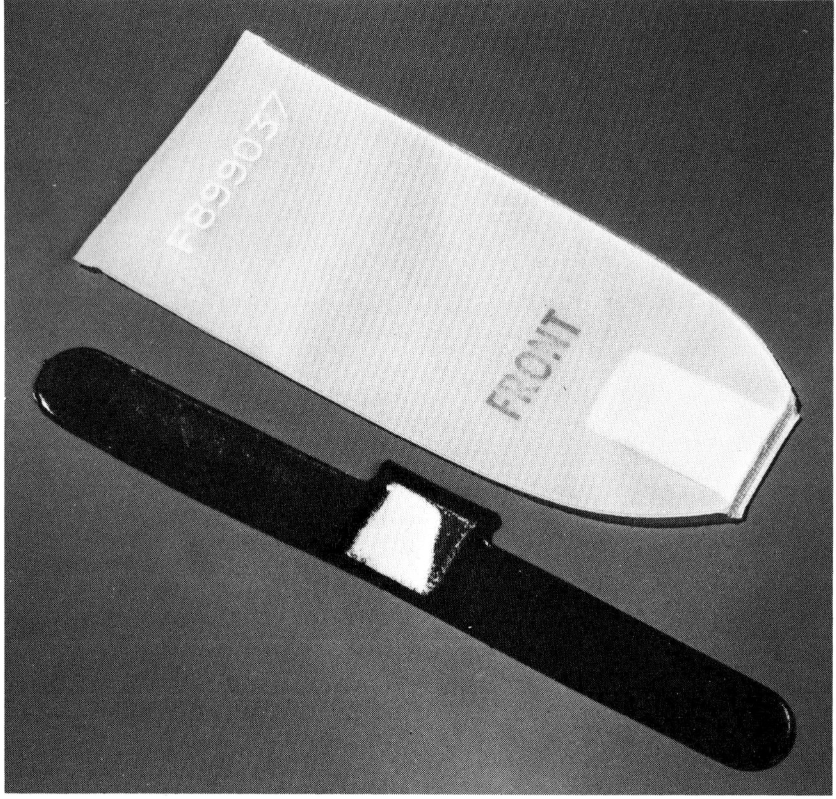

Figure 5.3 Dosemeters used for extremity monitoring. (Photograph courtesy of the National Radiological Protection Board, Harwell.)

A general summary of the recommended performance requirements for basic TL discriminating and non-discriminating individual dosemeters is given in table 5.1.

5.1.2 Environmental monitoring

Environmental monitoring is the measurement of radiation exposure from either cosmic and terrestrial radiation (natural background), or from the controlled or accidental release of man-made radioactivity.

Measurements of natural background radiation are important in the assessment of exposure to the population. They provide a data base upon which one can assess the relative magnitude of the effect of the

Table 5.1 Recommended performance requirements for TL individual dosemeters (CEC 1975b). Environmental effects are required to be insignificant.

Dosemeter	Absorbed dose range (Gy)	Photon energy range (keV)	Electron energy range (keV)	Quality information	Overall uncertainty	Precision 2σ at 10 mGy (%)	Photon energy dependence for body dose at 1000 mg cm^{-2} (%)	Photon angular response (%)	Fading at 25 °C (%)
Non-discriminating	2×10^{-4}–10	10–5×10^4	5×10^2–5×10^4	None required	⩽500 mGy −30%, + 50% or ± 0.5 mGy >500 mGy −20% +25%	⩽10	−20 +40	±30	<5
Discriminating	2×10^{-4}–10^2	10–5×10^4	5×10^2–5×10^4	Required over range 10–200 keV	⩽500 mGy −30%, + 50% or ± 0.5 mGy >500 mGy −20% +25%	⩽10	±15	±30	5

release of man-made radioactivity should it occur (figure 4.1). There is often a requirement to measure the natural background exposure in a specific area; for example, the measurement of cosmic radiation exposure in high-altitude commercial aircraft (Kramer *et al* 1977) and the measurement of the radioactivity of soil for TL dating (Aitken 1968). Routine measurements are carried out in the vicinity of nuclear power plants in order to detect exposure due to the controlled release of radioactivity. In the event of an accidental release, the measurements serve to assess the impact on the environment.

5.1.2.1 Environmental radiation. The major contribution to the radiation exposure rate at the Earth's surface (1 m is often used as a standard measurement height) is from gamma photon radiation. The exposure rate depends on the composition of the soil and rocks, the radioactive species present, the moisture content of the soil and the presence of surface water or snow. In a man-made environment the effects of building materials and shielding must also be considered. The contribution to the natural background from cosmic radiation varies with altitude, latitude, atmospheric conditions and solar activity. In contrast with terrestrial gamma radiation, cosmic radiation at the Earth's surface consists mainly of muons (approximately 80% of the total).

Although almost all the secondary radiations produced by cosmic radiation have energies greater than 100 keV (the energies of some secondary electrons generated by muon interaction can exceed 1 GeV), and the majority of terrestrial radiation comprises radiations having energies greater than 80 keV, there is evidence that there is a significant contribution from radiations with energies less than 50 keV.

5.1.2.2 Requirements for TL dosemeters for environmental monitoring. Even if environmental exposures of approximately 100 mR (2.58×10^{-5} C kg^{-1}) are assessed over one year, there is a need for a dosemeter which has a minimum detectable exposure of approximately 10 mR (2.58×10^{-6} C kg^{-1}). However if, as is often the case, there is a requirement to monitor continuously the environmental exposure (for example, to assess short-term fluctuations over a monitoring period of one month), then a minimum detectable exposure of approximately 1 mR (2.58×10^{-7} C kg^{-1}) is needed. It is important that the fading of stored signals should be small over the measurement period. The dosemeter should be unaffected by adverse environmental effects such as from humidity, high ambient temperature, high levels of ambient illumination (including natural ultraviolet radiation), etc. Neither the dosemeter nor its container should have a high enough natural radioactive mineral content to cause significant self-irradiation. For the calculation of absorbed dose or dose-equivalent the dosemeter should have approximate tissue-equivalence. Based on these and other considerations, the American National Standards Institute has published minimum performance criteria which environmental TL dosemeters should satisfy (ANSI 1975). Some of these criteria are summarised in table 5.2. In addition, the International Standards Organisation are currently preparing a standard for personal and environmental TL dosemeters (ISO 1979).

Table 5.2 Performance criteria for TLDs used for environmental monitoring (ANSI 1975).

Characteristics	Limits
Intra-batch uniformity—maximum deviation	±15%
Reproducibility of a single dosemeter—maximum deviation	±5%
Energy response	
80–3000 keV	≤20%
30–80 keV	≤200%
Effects of	
(i) ambient illumination	≤10%
(ii) humidity	≤10%
(iii) angular dependence	±10%
(iv) self-irradiation	$\not>$10 μR (2.58 × 10^{-9} C kg^{-1})h^{-1}
Overall measurement uncertainty	±30%

The performance of a particular dosemeter must of course be considered in relation to the performance of the system on which it is read out. The results of an intercomparison study (reported by Kramer *et al* 1977) of 20 different TLD systems have shown that, using the high-sensitivity phosphors CaF_2 and $CaSO_4$ whose properties are discussed in Chapter 3, exposures of less than 1 mR can be precisely determined. However, such are the potential uncertainties associated with individual dosemeters (e.g. in their calibration, irradiation, thermal treatment, etc), and the uncertainties associated with TL readers, that special calibration procedures must be applied to individual dosemeters. Results obtained in one laboratory cannot be universally applied. An example of a scheme for the calibration, exposure and measurement of TLDs for environmental measurements is illustrated in figure 5.4 (Piesch 1980).

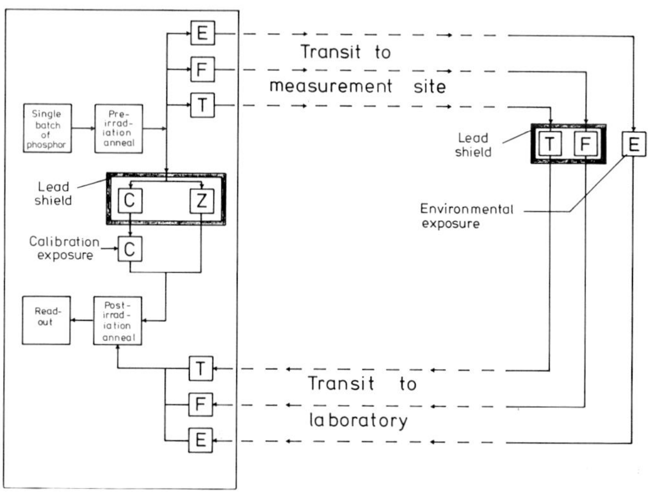

Figure 5.4 Schematic diagram illustrating the calibration, exposure and measurement of TLDs for environmental monitoring (after Piesch 1980). E, dosemeter for environmental measurements; F, pre-exposed dosemeters accompanying environmental dosemeters to correct for thermal fading; T, dosemeters accompanying environmental dosemeters to assess radiation exposure received in transit; C, calibration dosemeters stored in the laboratory (minimum of ten); Z, unexposed dosemeters to assess zero-dose readings (minimum of ten). The calibration exposure should be of the order of 1000 times the minimum detectable exposure (calculated from three times the standard deviation on a set of zero-exposure readings).

5.2 Charged Particle and Neutron Dosimetry

5.2.1 TL response to high LET charged particles

There have been many studies of the TL responses of a number of phosphors as a function of LET of the incident radiation. The measurements have been most often performed on phosphors irradiated with alpha particles, but protons and heavy ions have also been used. For lithium fluoride at least, the results of these studies have shown a general agreement that the TL response of this phosphor decreases with increasing LET, as shown in table 5.3. As the integrated values of TL listed for LiF include the area under the higher-temperature peak 6 (285 °C) as well as peak 5 (210 °C), they are relatively higher than the values based on the height of peak 5 alone. The relative importance of peaks 5 and 6 in the measurement of integrated TL has been observed to depend on the LET of radiation and the isotopic enrichment of the phosphor. Driscoll (1978) found that, for both TLD 600 and 700 powder phosphors ($< 26\ \mu$m grain size) irradiated with a given absorbed dose, the ratio of peak 5 (high LET radiation, 5.5 MeV alpha) to peak 5 (low LET radiation, ^{137}Cs gamma) was constant (0.3). However, for peak 6 the corresponding ratios were 0.3 for TLD 700 and 1.8 for TLD 600. Lithium borate, with its relatively simple glow curve, does not show any such effect. For example, there is close agreement between the integrated and peak height measurements obtained by Mäjborn *et al* (1977).

While the general trend of results for lithium fluoride shown in table 5.3 is of decreasing TL response with increasing LET in the range 100–1000 keV μm^{-1}, two sets of results do not fit the pattern. The TL responses for LiF obtained by Tochilin *et al* (1968) show a plateau between 100 and 1000 keV μm^{-1}, and Henson and Thomas (1978) measured an even higher value, 0.52 at 200 keV μm^{-1}. The measurements of Tochilin *et al* for lithium borate are also high: 1.0 between 100 and 1000 keV μm^{-1}, compared with 0.3–0.2 between 260 and 321 keV μm^{-1} obtained by Mäjborn *et al* (1977) and 0.37 at 263 keV μm^{-1} obtained by Lakshmanan and Ayyangar (1976). There does not appear to be a satisfactory explanation for these discrepancies. Horowitz *et al* (1979a) have listed a number of effects, variations in which could affect the precision of measurement of TL response. These include annealing methods, cooling rates, spectral variations in TL emission, readout heating rates, and TL self-absorption. However, their measurements on LiF (TLD 100, 600 and 700) in which all experimental procedures were standardised revealed marked differences in the TL responses of the different batches as shown in table 5.3. It is argued that these differences are a result of variabilities in composition between batches of phosphor; in particular, the amount of barium present was found to vary between 1 and 30 ppm.

It has been noted that for LiF supralinearity and sensitisation are both related to the LET of the radiation (Suntharalingam and Cameron 1969). Supralinearity is much less for high LET than for low LET radiation. Sensitisation of the phosphor following a large absorbed dose is less for subsequent exposure to high LET than to low LET radiation.

Table 5.3 TL responses of LiF, $Li_2B_4O_7$:Mn and BeO for high LET particle radiation. TL response is expressed as integrated (area) TL denoted by 'A', or as peak height 'P'.

Phosphor	Particle energy (MeV)	Average $\overline{\text{LET}}_\infty$ (keV μm^{-1})	TL response ($^{60}Co \equiv 1.0$)	References
LiF	^1H, He, ^{12}C, ^{20}Ne, ^{40}Ar	0.25–2	1.05–1.2 P	Tochilin *et al* (1968)
		2–10	1.2–0.9 P	
		10–100	0.9–0.4 P	
		100–1000	0.4 P	
	5.5α	242	0.135 P	Lucas and Rainbolt (1968)
	4.48α	263	0.126 P	
	3.38α	291	0.088 P	
	2.09α	318	0.066 P	
	4.48α	263	0.234 A	Harvey and Townsend (1971)
	3.86α	280	0.245 A	
	2.85α	301	0.209 A	
	1.68α	321	0.130 A	
	13.3 ^1H	21	0.86 A	Jahnert (1972)
	2.4 ^1H	57	0.56 A	
	3.7–1.7α	305–340	0.165 A	
	4.7α	260	A 0.21 0.11 P	Mäjborn *et al* (1977)
	3.8α	281	A 0.22 0.09 P	
	2.8α	304	A 0.18 0.07 P	
	1.7α	321	A 0.15 0.06 P	

	$^{18}Ar^{+}$	200	0.52 A	Henson and Thomas (1978)
	5.7α	238	A 0.23 0.16 P	Bartlett and Edwards (1979)
	4.9α	255	A 0.25 0.15 P	
	4.3α	269	A 0.22 0.13 P	
	2.5α	309	A 0.17 0.09 P	
	3.8α	295	A 0.17 (TLD 100)	Horowitz *et al* (1979a)
			A 0.21 (TLD 600)	
			A 0.29 (TLD 700)	
$Li_2B_4O_7$:Mn	^{1}H, He, ^{12}C, ^{20}Ne, ^{40}Ar	0.25–10	1.05–1.25 P	Tochilin *et al* (1968)
		10–100	1.25–1.1 P	
		100–1000	1.1 –0.9 P	
	4.7α	260	A 0.31 0.32 P	Mäjborn *et al* (1977)
	3.8α	281	A 0.31 0.32 P	
	2.8α	304	A 0.29 0.30 P	
	1.7α	321	A 0.20 0.20 P	
BeO	^{1}H, He, ^{12}C, ^{20}Ne, ^{40}Ar	0.25–10	1.05–2.0 P	Tochilin *et al* (1968)
		10–100	2.0 –2.1 P	
		100–1000	2.1 –2.2 P	

Attix (1975) has considered these effects in LiF on the basis of the track interaction model as follows. The average separation between ionising events along the track of a high LET charged particle (e.g. 2 MeV alpha) is approximately 0.14 nm, which is much less than the distance separating nearest-neighbour trapping centres (6.8 nm). Therefore it is highly unlikely that more than about 2% of electron–hole pairs produced as a result of ionisation events will become trapped; the vast majority will recombine immediately. Although the separation between F centres (which play an important role along with Ti^{4+} ions in the TL recombination and emission process) is reduced along high LET tracks, the low trapping efficiency predominates. On the basis of this model it is to be expected that LiF will have a low TL response to high LET radiation. Further, at relatively low levels of absorbed dose only intra-track TL emissive recombinations of electrons and holes will be possible as the inter-track separation distances are large compared with the electron–hole migration distances. At high levels of absorbed dose from low LET radiations the tracks are packed much closer together, thus enabling inter-track recombinations, and resulting in the onset of supralinearity. However, for high LET radiations, very much higher levels of absorbed dose are required to produce the density of tracks where the inter-track distances became comparable with the intra-track recombination distances. Therefore supralinearity does not occur.

5.2.2 Neutron dosimetry

Neutrons are uncharged particles and consequently produce negligible ionisation in matter. Their detection and measurement depend on the absorption of secondary radiation (produced as a result of their interactions with nuclei) by either of the following methods.

(i) Capture by a nucleus resulting in (*a*) prompt emission of a fast charged particle, or (*b*) fission of the nucleus and formation of fission fragments, or (*c*) formation of a radioactive nuclide.

(ii) Scatter by a nucleus, e.g. hydrogen (proton), which recoils producing ionisation.

5.2.2.1 Thermal neutrons. Neutrons whose energies are in equilibrium with those of the atoms or molecules of the surrounding matter are termed thermal neutrons. This condition results from multiple scattering in a material (moderator) whose scatter cross section is much greater than its absorption cross section. The distribution of velocities of thermal neutrons may be represented by a Maxwellian distribution with a most probable velocity at normal ambient temperature corresponding to an energy of approximately 0.025 eV. However, all neutrons of energies less than 0.5 eV are generally referred to as thermal neutrons.

The thermal neutron response of a thermoluminescence phosphor depends on the capture cross sections of all of its constituent atoms.

Table 5.4 Thermal neutron responses of some TLD phosphors (Ayyangar *et al* 1974).

TL phosphor	Thermal neutron response equivalent, R ^{60}Co $(10^{10}\ neutrons/cm^2)^{-1}$
LiF:Mg:Ti	
TLD 600	870–2190
TLD 100	65–535
TLD 700	0.7–2.5
$Li_2B_4O_7$:Mn	230–500
$CaSO_4$:Mn:^{6}Li	1050
$CaSO_4$:Dy	0.38–0.52
$CaSO_4$:Tm	0.21–0.23
CaF_2:Mn	0.1–0.6
CaF_2:Dy (TLD 200)	0.5–0.65
CaF_2 (fluorites)	0.16

The responses of many TL phosphors have been measured (see table 5.4). Unfortunately because of differences in readout conditions, uncertainties and variations in the relative proportions of the nuclear constituents of phosphors, differences in sample thickness resulting in varying degrees of self-shielding, etc, the data have not always been in close agreement; the ranges indicated in table 5.3 represent the limits of the spread on the reported data for these phosphors. The phosphors which have been shown to have the highest thermal neutron responses are LiF : Mg : Ti as TLD 600 and TLD 100, $Li_2B_4O_7$: Mn and $CaSO_4$: Mn : ^{6}Li.

Because thermal neutrons are invariably accompanied by photon radiation it is also important to have phosphors with a low thermal neutron sensitivity, such as LiF : Mg : Ti (TLD 700), $CaSO_4$: Dy and $CaSO_4$: Tm, in order to make mixed-field dosimetry measurements.

In principle, there are two methods by which one can differentiate between thermal neutron and photon doses.

(i) A subtraction method can be used, involving the use of two detectors in the field—one with a high thermal neutron response (e.g. TLD 600), and the other with a low response (e.g. TLD 700). Both detectors have a similar photon response. In principle, by calibration and subtraction of the responses, both thermal neutron fluence and photon absorbed dose may be calculated.

(ii) Differences can be noted in the TL characteristics of a phosphor exposed to neutrons compared with a nominally identical one exposed to photons. Such differences may, for example, be in the TL emission spectrum of a phosphor, the onset of supralinearity, or in the glow curve structure. This last effect has been used in LiF:Mg:Ti where the 285 °C peak is considerably enhanced relative to the 210 °C main dosimetry peak due to the absorption of the high LET α and ${}^{3}_{1}H$ particles produced as a result of the ${}^{6}_{3}Li\ (n,\alpha)\ {}^{3}_{1}H$ reaction. By measuring the ratio of the heights of the peaks, the neutron and photon doses can be separated. However, mixed-field dosimetry is complex in practice and subject to many potential sources of uncertainty. A basic premise is that the neutron and photon exposures are non-interactive, i.e. neither alters the intrinsic response of the phosphor. This is not at all certain to be the case for all phosphors. Mason (1970) reported that the exposure of LiF:Mg:Ti to neutrons affected its subsequent photon response by shifting the onset of supralinearity, whereas Blum *et al* (1973) found no interaction for CaF_2:Mn.

5.2.2.2 Intermediate, fast and relativistic neutrons. Intermediate neutrons have energies ranging from 0.5 eV to 10 keV, fast neutrons energies ranging from 10 keV to 10 MeV, and relativistic neutrons energies greater than 10 MeV. In these energy ranges the TL process can in principle be due to the absorption of energy from the following:

(1) recoil light nuclei such as protons resulting from scatter interactions with neutrons—even if the phosphor is deficient in light nuclei the effect can be enhanced by the use of a slab of scattering material (radiator);

(2) fission fragments produced as a result of fission interactions between the neutrons and foils of fissionable materials;

(3) ionising radiation produced by nuclear reactions between neutrons thermalised by the moderating properties of the dosemeter and nuclei in the phosphor—where a moderator material is located behind the phosphor an albedo dosemeter is formed;

(4) ionising radiation emitted by radioactive decay of radioisotopes produced by neutron activation of the phosphor elements;

(5) gamma photons emitted by excited target nuclei from an inelastic scattering with relativistic neutrons.

Let us now examine in some detail the neutron response and mixed-field dosimetry characteristics of some phosphors.

LiF : Mg : Ti is commercially available as TLD 600, TLD 700 and TLD 100 containing different amounts of ^{6}Li and ^{7}Li isotopes, as shown in table 3.2. The relatively high thermal neutron responses of TLD 600 and TLD 100 compared with that of TLD 700 (table 5.4) are due to the high absorption cross section (945 b) of the ^{6}Li nucleus resulting in the reaction $^6_3$Li (n,α) $^3_1$H. This reaction produces a 2.07 MeV alpha particle and a 2.74 MeV triton whose absorptions by the phosphor result in the TL signal. This reaction dominates the neutron response of all phosphors containing ^{6}Li. The variations in neutron responses of TLD 600 and TLD 700 are shown in figure 5.5 as a function of neutron energy. The theoretical kerma response data of Horowitz and Freeman (1978) are calculated by weighting the energies of the charged particles produced as a result of all the significant neutron-induced reactions, with the TL–LET dependence of the particles.

The experimental data in figure 5.5 are those of Tanaka and Furuta (1977). At 100 keV and 1 MeV neutron energies, the ratios of the responses of TLD 600 and TLD 700 are approximately two and one order of magnitude, respectively. The equivalent response of TLD 600 and TLD 700 to photons coupled with their marked differential response to neutrons has led to their use as paired dosemeters in mixed neutron–photon fields (e.g. Furuta and Tanaka 1972, Tanaka and Furuta 1974, 1977).

Recently, however, Horowitz *et al* (1979b) have summarised the limitations of this method: the comparatively high thermal-neutron-induced TL signal; its variability with the angular-dependent effect of self-shielding; and the uncertainty in the neutron-induced TL response of dosemeters due to batch to batch variations. At high neutron flux levels (5×10^9 neutrons/cm^2), Piesch *et al* (1978) have observed a decrease in the TL response of TLD 600 due to radiation-induced damage. Also the β-decay of the tritium produced as a result of the $^6_3$Li (n,α) $^3_1$H reaction results in a self-irradiating effect which produces a build-up in the zero absorbed dose signal.

The peak ratio (210/285) in LiF has been used for mixed-field dosimetry by a number of people (e.g. Ayyangar *et al* 1968, Mason 1970, Busuoli *et al* 1970, Marshall *et al* 1977). The ratios of the peaks for TLD 600 phosphor for photons of a range of energies and for thermal neutrons are shown in table 5.5. Using the calibrated responses of the two peaks for photon and neutron exposure, and the measured values of both peaks for a mixed-field irradiation, the neutron dose can be calculated by the solution of two simultaneous equations. Using a two-temperature (240 and 300 °C) readout cycle, Marshall *et al* (1977) measured a thermal-neutron flux of 2×10^6 neutrons/cm^2 with a standard deviation of $\pm 10\%$ in the presence of 10 mGy of gamma rays.

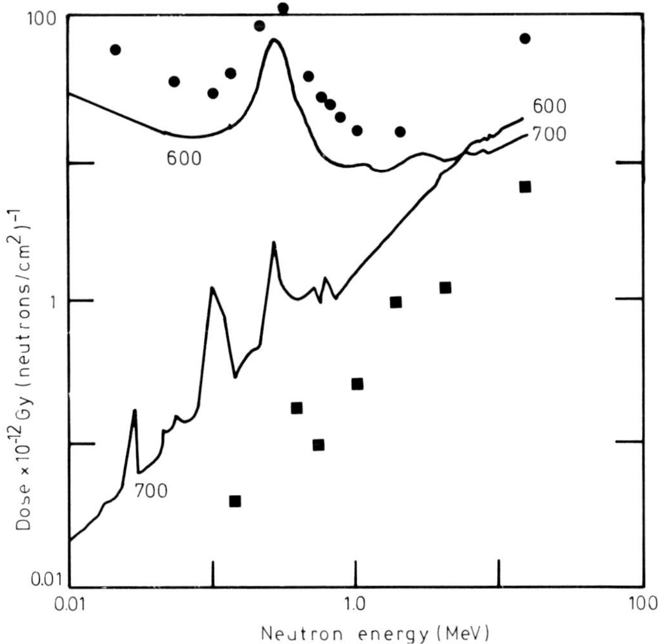

Figure 5.5 TL responses of LiF, TLD 700 and TLD 600 as functions of neutron energy. Full curves, theoretical values of Horowitz and Freeman (1978). ■ (TLD 700) and ● (TLD 600), experimental data of Tanaka and Furuta (1977). (Horowitz and Freeman 1978, reprinted with the permission of the North-Holland Publishing Co.)

The glow peak structure of $Li_2B_4O_7$: Mn is identical for both photon and neutron exposures (Lakshmanan *et al* 1976) and the usefulness of the phosphor may lie in producing ^{6}Li and ^{10}B enchanced types.

Blum *et al* (1973) investigated the mixed-field response of CaF_2 : Mn : PTFE disc dosemeters irradiated in a cyclotron fast-neutron beam. The non-uniform photon energy response of CaF_2:Mn was not important since the photon absorbed dose was a small fraction of the total. Dosemeters were exposed simultaneously in pairs, each member of a pair having either a hydrogenous (polythene) radiator or a non-hydrogenous (lead) screen. The results of both phantom and *in vivo* measurements indicated that the dosemeters generally underestimated the depth absorbed dose. This was mainly due to the non-uniform

Table 5.5 Ratio of integrated peak area for 210 and 285 °C peaks in TLD 600 for photons of different energies and thermal neutrons (Marshall *et al* 1977).

Type of radiation	Ratio 210/285 °C peaks
^{226}Ra photons	17.9
X-rays (keV)	
148	11.5
109	10.3
87	9.3
61	9.6
48	8.9
30	10.1
Thermal neutrons	1.8

directional response of the discs when scattered neutrons were not incident normally on the surface of the dosemeter. This angular dependence of the proton radiator–TLD style dosemeter can be overcome by forming intimate mixtures of the TL phosphor and the hydrogenous radiator material. The radiator material must then be able to withstand the high-temperature conditions associated with the readout and anneal of dosemeters, or be separable from the TL phosphor prior to readout and anneal. Both liquid (usually alcohol; see Wingate *et al* 1965) and solid forms of homogeneous radiators have been developed. Solid forms are generally more convenient to use.

Blum *et al* (1972) developed a paired dosemeter consisting of two small PTFE tubes, one containing an intimate mix of one part $CaSO_4$:Tm powder and three parts glucose powder, and the other containing only $CaSO_4$:Tm powder. After simultaneous exposure of both tubes to a fast-neutron beam the glucose was removed by washing and drying. Both sets of $CaSO_4$:Tm powder were then read out. A differential response of approximately 4:1 was obtained, providing sufficient discrimination between fast neutrons and photons. The maximum non-uniformity in angular response was less than 10%.

Becker *et al* (1973) used the organic hydrogenous material p-sexiphenyl (5.68% hydrogen) as a high-melting-point, stable homogeneous radiator. The neutron-sensitive dosemeter was formed from a hot pressed mixture of p-sexiphenyl powder with $CaSO_4$:Dy powder in the ratio 1:1 at 200°C and a pressure of 2000 kg cm^{-2}. Similar non-neutron-sensitive dosemeters were made by substituting PTFE for the p-sexiphenyl. Table 5.6 shows the ratio of the responses of the two forms of dosemeter to 14 MeV and fission neutrons. However the p-sexiphenyl dosemeters have a comparatively high TL response to UV radiation necessitating their use and handling under red light. Instead of incorporating a proton radiator material in the dosemeter a moderator material can be used. Such a system moderates fast neutrons to thermal energies allowing their detection by paired high and low thermal-neutron-sensitive phosphors. Figure 5.6 shows the net thermalised-neutron response (TLD 600 minus TLD 700) for pairs of such phosphors at the centre of a 10 inch diameter polythene moderator sphere and irradiated with neutrons of various energies. The experimentally derived centre to surface ratio is also shown as a function of neutron-effective energy. The neutron-effective energy can be assessed by measurement of this ratio and the neutron dose thus established.

Table 5.6 Fast-neutron responses of $CaSO_4$:Dy:p-sexiphenyl (polythene-shielded) compared with $CaSO_4$:Dy:PTFE (PTFE-shielded) for 14 MeV and fission spectrum neutrons (Becker *et al* 1973).

Neutrons	Ratio of p-sexiphenyl/PTFE responses	Neutron/photon dose
Fission spectrum	3:1	7:1
14 MeV	10:1	25:1

An albedo dosemeter measures the thermal-neutron flux which results from the scattering and moderating effect of the body on incident fast neutrons. The designs of albedo dosemeters are varied and numerous, but their basic principle of operation is illustrated in figure 5.7.

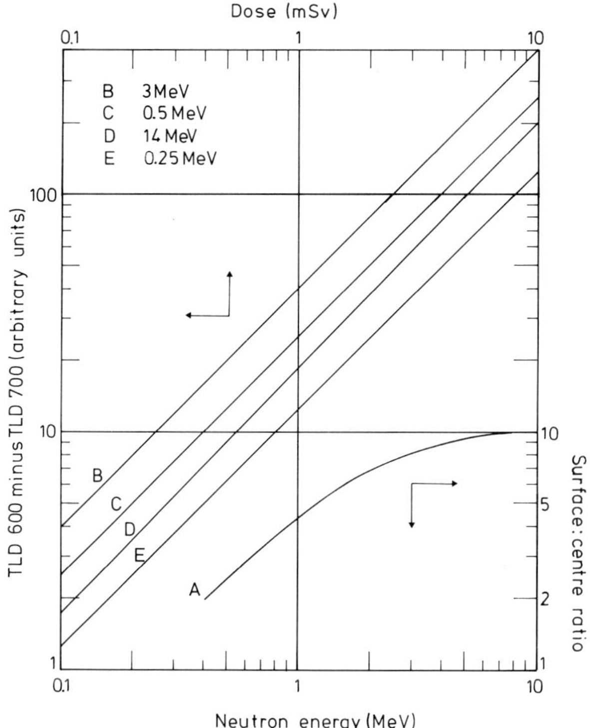

Figure 5.6 A, experimental centre to surface ratio for a 10 inch diameter polyethylene sphere as a function of neutron energy. B–E, TLD 600 minus TLD 700 thermoluminescence–dose response at the centre of a sphere, as a function of neutron energy. (Engelke and Israel 1974, reprinted with the permission of Pergamon Press Ltd, Oxford.)

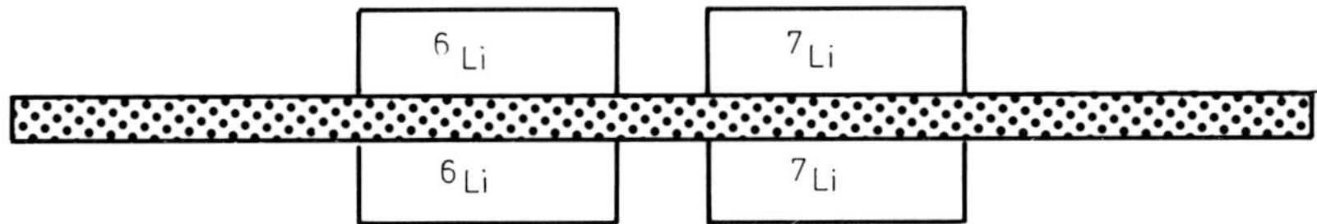

Figure 5.7 Basic design of an albedo dosemeter. The bottom pair of ^{6}Li, ^{7}Li dosemeters measure the albedo neutrons; the top pair measure the incident thermal neutrons. The cadmium- or boron-loaded plastic shield (shaded) prevents incident neutrons reaching the lower pair of dosemeters. The shield should extend far enough effectively to prevent thermal neutrons entering the body at a position where they may be scattered and detected as though they were albedo neutrons. (Griffith *et al* 1979, reprinted with the permission of Pergamon Press Ltd, Oxford.)

The top pair of phosphors measures the incident thermal-neutron flux (^{6}Li minus ^{7}Li) while the bottom pair measures the albedo thermal-neutron fluence and, by calibration, the fast-neutron dose, provided

(1) that the thermal neutron absorber, which is generally cadmium or ^{10}B-loaded plastic, extends sufficiently far on both sides between the pair of phosphors to ensure that few incident thermal neutrons are detected by the bottom phosphors—in practice, an experimentally derived correction factor is applied to the readings to allow for this—and

(2) that in a mixed neutron–photon field corrections are made for the photon absorption of the cadmium foil (particularly for photons of energies < 100 keV).

Any incident photon dose may be measured with the ^{7}Li phosphor provided that no thermal neutrons are present in the incident beam. Thermal neutrons react with the cadmium to produce γ-rays which contribute to the response of the phosphors. Many types of albedo dosemeter have been designed, but it is beyond the scope of this text to discuss their characteristics in detail. A useful review of their development, characteristics and use for personal neutron dosimetry has been written by Griffith *et al* (1979).

5.3 Thermoluminescence Dating

5.3.1 Introduction

The principle on which the techniques of thermoluminescence dating are based is that if, in the distant past, a material has been heated it will at that instant have given up all of its stored thermoluminescence. In the case of pottery, for example, the firing of the clay by the potter would perform this function. Any thermoluminescence which the material now displays when heated will have resulted from irradiation of the material during the intervening period. Assuming that no artificial irradiation of the material has occurred, all of its TL is due to the absorbed dose from the radiations emitted by naturally occurring radioisotopes, and to a much lesser extent from cosmic radiation. For example, in the case of pottery which has been buried in the ground the sources of radiation are those radioactive elements belonging to the uranium and thorium series, and potassium (^{40}K) which are present within the clay matrix. In addition there will be a contribution from β and γ

radiation originating in the surrounding soil, and also a relatively small contribution from cosmic radiation.

This so-called natural TL (NTL) can be expressed as

$$(\text{NTL}) = SD, \tag{5.1}$$

where S is the thermoluminescence sensitivity of the material per unit absorbed dose (at this stage it is assumed that S is a linear function of absorbed dose although as we shall see this is not always the case), and $\dot{D}$ is the total absorbed dose.

If the total time during which the pottery has been irradiated since firing is T years, then

$$D = T\dot{D}, \tag{5.2}$$

where $\dot{D}$ is the total annual effective absorbed dose rate, and hence

$$(\text{NTL}) = ST\dot{D}$$

or

$$T = \frac{(\text{NTL})}{SD}. \tag{5.3}$$

$\dot{D}$ consists of contributions from α, β and γ radiations and a few per cent cosmic radiation, $\dot{D}_\alpha$, $\dot{D}_\beta$, $\dot{D}_\gamma$ and $\dot{D}_c$, respectively.

However, the thermoluminescence sensitivity for α radiation, S_α, is a fraction of that for β, γ and cosmic radiations, S_β. Hence equation (5.3) may be re-written as

$$T = \frac{(\text{NTL})}{S_\alpha \dot{D}_\alpha + S_\beta(D_\beta + D_\gamma + D_c)}. \tag{5.4}$$

By defining a factor

$$k = \frac{S_\alpha}{S_\beta} \tag{5.5}$$

we obtain

$$T = \frac{(\text{NTL})/S_\beta}{k\dot{D}_\alpha + \dot{D}_\beta + \dot{D}_\gamma + \dot{D}_c}. \tag{5.6}$$

The quantity (NTL/S_β) is equivalent to that artificial β absorbed dose which will produce a TL signal equivalent to the natural thermoluminescence (NTL).

In principle, therefore, the evaluation of the age (T) of the pottery is straightforward. The values of $\dot{D}_\alpha$, $\dot{D}_\beta$ and $\dot{D}_\gamma$ can be obtained by analysis of the radioactive content of the pottery and surrounding soil. $\dot{D}_c$, which amounts to only a few per cent of the total, can be calculated. In practice, however, there are many factors which can affect the precision of the measurement and very careful experimental techniques and precise calculations have to be performed in order to achieve meaningful results. For example, the moisture content of the soil and pottery is important in absorbed dose calculations and the escape of radon and thoron gases will alter the absorbed dose considerably.

Pottery is not a homogeneous material in either a physical or radioactive sense. It consists of a matrix of clay particles, typically $< 10\ \mu m$ diameter, in which are embedded many quartz and a few zircon inclusions up to a few millimetres in size. The clay matrix contains the bulk of the radioactivity in the pottery (typically 3 ppm uranium, 12 ppm thorium and 2 ppm potassium) and as a result the very small clay matrix grains are uniformly irradiated with α, β and γ radiation. However the clay grains have a very low thermoluminescence sensitivity. The quartz inclusions on the other hand have a relatively high thermoluminescence sensitivity but effectively contain no radioactive material. Their thermoluminescence is therefore derived from absorbed dose due to α, β and γ radiations from the clay matrix and γ radiations (within approximately 30 cm) from the surrounding soil. Because of the limited maximum penetration of α radiation (typically 25–40 μm) and β radiation (typically 500 μm) into the quartz particles, the absorbed dose distribution within the particles is non-uniform and will depend on particle size. The zircon inclusions are few in number but contain relatively high concentrations of natural radioactive elements (typically 50–300 ppm uranium). Their thermoluminescence sensitivity is generally high but can vary considerably from grain to grain.

By separation of the various grains described above, three different methods have been developed to determine the age of pottery.

(i) *Fine-grain method.* This method involves the isolation of fine clay grains (diameters $< 8\ \mu m$). It may be assumed that the absorbed dose to these grains is uniform and consists of contributions from α, β and γ radiations. Both α and β artificial irradiations are therefore required. Experimental samples are prepared by depositing a thin layer of grains onto aluminium discs. The uranium and thorium contributions to the absorbed dose are measured by α counting and the potassium by chemical analysis of its concentration (Zimmerman 1971).

(ii) *Quartz inclusion method.* This method consists of selecting quartz grains large (typically 100–150 μm) compared with the maximum range of α radiation. As the quartz grains contain essentially no radioactivity, their total α absorbed dose is due entirely to α radiation from the surrounding clay matrix. This absorbed dose is deposited in the surface layer of the quartz grain. Removing this surface layer by etching with hydrofluoric acid leaves a quartz grain whose absorbed dose is due entirely to β and γ radiations, so that no α dosimetry is required. In practice, however, a correction has to be applied to take account of attenuation of β radiation (Fleming 1970).

(iii) *Zircon inclusion method* (Sutton and Zimmerman 1976). In contrast with quartz, zircon inclusions contain a relatively high concentration of natural radioactivity and as a result the effective absorbed dose deposited in a grain is large. This is so large that the contribution to absorbed dose from the surrounding soil can be ignored, and soil samples are therefore not necessary for absorbed dose evaluations. However, because thermoluminescence sensitivity and radioactive content vary greatly from grain to grain, the measurement technique is applied to single grains. Even individual grains show a marked spatial non-uniformity of cathodoluminescence intensity and uranium content, and there is also a strong anti-correlation between the two. Grains are therefore selected on the basis of their spatial uniformity of cathodoluminescence. As the total absorbed dose within a grain is due almost entirely to α radiation originating in the grain, it is important to measure its uranium and thorium contents accurately. This is done by irradiating grains with neutrons and 30 MeV ^{3}He ions and comparing the numbers of fission tracks produced with those produced in similarly irradiated glasses containing known concentrations of uranium and thorium.

5.3.2 Supralinearity and sensitisation

The TL signal from a pottery material does not increase linearly with increasing absorbed dose. At high levels of absorbed dose the response of the material may be markedly supralinear, and saturation will eventually occur. Also, after a material has received a large absorbed dose and has been read out, its subsequent TL sensitivity is increased (sensitisation). These two effects combine to make the determination of natural absorbed dose difficult, but a method has been developed which corrects for these effects. By preparing several samples of the material and irradiating them to different values of absorbed dose (artificial absorbed dose), a TL–absorbed dose response curve can be drawn (figure 5.8, curve A). The linear extrapolation of this curve will intercept the absorbed dose axis at a value which would correspond to the natural absorbed dose (NAD) if the TL–absorbed dose curve were linear. A correction for supralinearity is obtained by carrying out irradiations on a series of similar samples which have first been read out and therefore emptied of their natural TL; this gives curve B. The extrapolation of the

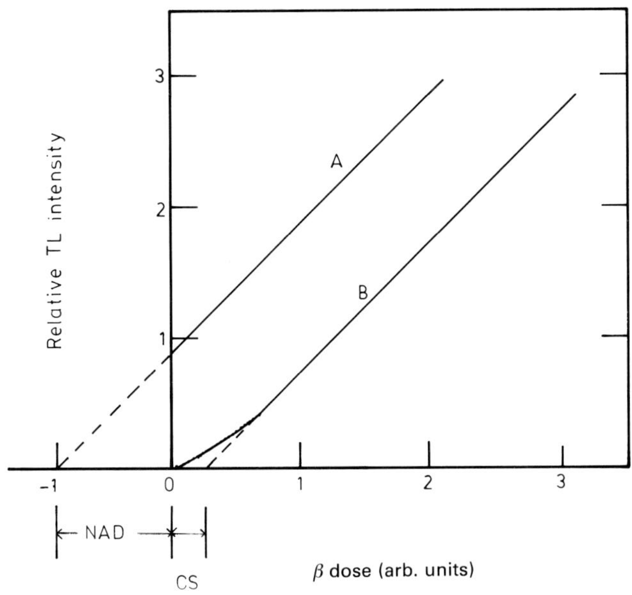

Figure 5.8 TL–absorbed dose response (curve A) for a sample of pottery. Curve B, correction for superlinearity (CS).

linear portion of this curve intercepts the absorbed dose axis at the supralinear correction value (CS). The basic assumption in this procedure is that despite the change in TL sensitivity, which is known to occur due to heating, the supralinearity correction for both curves is the same. The good correlation between the ages obtained by this method and the known ages of certain pottery suggests that this assumption is valid.

5.3.3 Fading effects

The glow curves associated with natural thermoluminescence do not contain any low-temperature glow peaks, as illustrated in figure 5.9. This is to be expected as the traps associated with these glow peaks will not have retained their trapped electrons after the many centuries of burial. Artificial irradiation of the pottery will produce a glow curve with the same high-temperature peak structure but, in addition, lower-temperature peaks, as shown in figure 5.9. Thermoluminescence measurements cannot be made on a part of the glow curve in which

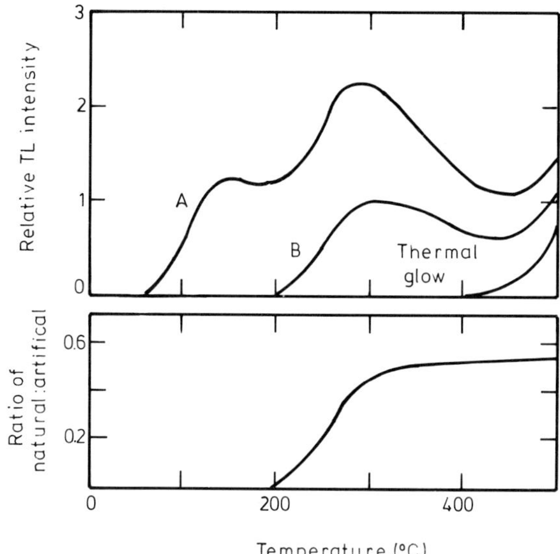

Figure 5.9 TL glow curves associated with A, artificial β radiation, and B, natural irradiation of pottery. The ratio of natural:artificial TL is used to determine the temperature above which no fading has taken place (plateau test).

fading is suspected. It is therefore necessary to select a minimum temperature below which the TL will not be recorded. This temperature is found by measuring the ratio of the natural TL (NTL) to the artificial TL (ATL) as a function of glow temperature. For those glow temperatures at which negligible fading of natural TL has occurred the ratio will be a constant. The plot of the ratio against temperature will give a plateau, as illustrated in figure 5.9; consequently this test is called the plateau test. Glow peaks in the temperature range 300–500 °C are normally used for TL dating. This range of temperatures is higher than that normally used with TLD phosphors, and ordinary commercial TLD readers are not suitable for readout. In particular the background incandescence radiation is much greater, so that special infrared absorbing filters have to be used. Some minerals, notably feldspars, exhibit so-called anomalous fading of the high-temperature dating glow peaks (300–500 °C). The observed rate of fading of stored TL cannot be explained in terms of the probability of thermally stimulated release of electrons predicted by the Boltzmann equation as discussed in § 3.1.3.1. The escape of electrons is most probably due to tunnelling.

5.3.4 Pre-dose method

In the dating of quartz the glow peak normally used is around 375 °C, although a test irradiation of a few tens of mGy reveals the presence of a low-temperature glow peak at 110 °C. An interesting effect is observed with this peak, which forms the basis of a dating technique known as the pre-dose method (Fleming 1973, Aitken and Murray 1976).

A sample of naturally irradiated quartz is given a test absorbed dose of between 10 and 100 mGy (insignificant compared with the natural absorbed dose), and the TL sensitivity (S_0) of the 110 °C peak is measured. The sample is then heated to 500 °C (emptying the natural TL) and the TL sensitivity (S_n) of the 110 °C peak is measured. The sample is again heated to 500 °C and the sensitivity (S'_n) of the 110 °C peak again measured. The sample is then given a calibration absorbed dose of several Gy of β radiation (β Gy), heated to 500 °C and the sensitivity of the 110 °C peak ($S'_{n+\beta}$) measured. It is found that

$$S_n > S_0, \quad S_n \div S'_n \quad \text{and} \quad S'_{n+\beta} > S'_n.$$

Since $S_n \div S'_n$, no increase in sensitivity is expected due to heating alone. $(S_n - S_0)$ is proportional to the natural absorbed dose (NAD) and $(S'_{n+\beta} - S'_n)$ is proportional to the calibration absorbed dose β Gy. Hence

$$(\text{NAD}) = \frac{(S_n - S_0)}{(S'_{n+\beta} - S'_n)} \times \beta\,\text{Gy}.$$

The sensitivity enhancement saturates at around 5 Gy, thus limiting the method to samples of ages of a thousand years or so.

5.3.5 Phototransferred thermoluminescence (PTTL)—transfer dating

As with certain TLD phosphors (see § 2.7), certain minerals exhibit PTTL following irradiation with ultraviolet radiation; this results in the transfer of electrons from very deep traps to those corresponding to low-temperature glow peaks (Bailiff *et al* 1977). The technique is particularly useful in those samples which exhibit anomalous fading. The measurement of high-temperature TL peaks against a high-incandescence background is also avoided.

6 TLD Instrumentation

6.1 Introduction

TLD readers can be classified into three categories:

(1) research designated readers,
(2) commercial readers,
(3) purpose-built automated systems.

While the same general design principles apply to all readers the specific details vary considerably according to the requirements of the particular application.

Research readers are generally built 'in house' and are as varied in design as they are numerous. The commercial readers are of most interest to the average TLD user. Many commercial systems are now available and they fall into two types: those specific to a particular design of dosemeter, and those which can accept a variety of dosemeter forms or phosphor types. They may be manual or semi- or fully automated systems. Many commercial readers are also suitable for research purposes, and some offer research options such as versatile heating programmes. The purpose-built automated systems are few in number and are very specific in application. They are often the result of 'in house' development by a large organisation requiring a high-throughput personal dosimetry system designated to one type of dosemeter.

The essential features common to every TLD reader are (see figure 6.1):

(1) a phosphor heating system,
(2) a light collection and detection system,
(3) a signal-measuring system,
(4) a display and recording system.

6.2 Heating the Dosemeter

As we have already seen many different physical forms of dosemeter are used in thermoluminescence dosimetry, e.g. loose powder, solid extrusions of various shapes and sizes, and powder incorporated in a binder material such as PTFE. These different forms vary considerably in their thermal properties. An ideal dosemeter would have a flat, uniform surface providing excellent thermal contact with the heat source, high thermal conductivity and low thermal capacity to eliminate thermal gradients in the phosphor. Thermal gradients cause low resolution in the glow peak structure and possible uncertainty in the integrated signal. The presence of large 'thermal sinks' at or around the dosemeter should be avoided as they result in long high-temperature tails on the glow curve.

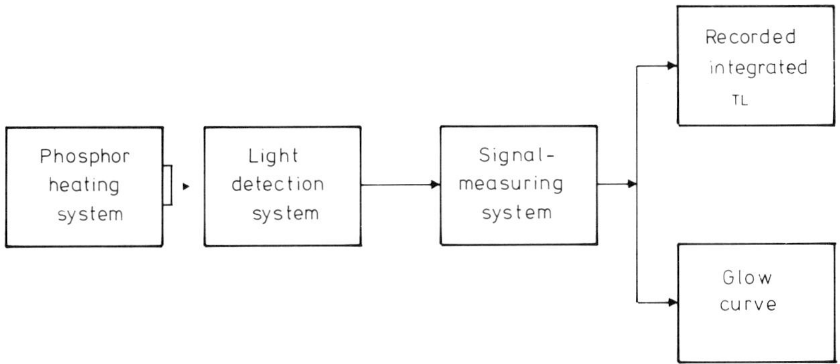

Figure 6.1 Schematic diagram illustrating features common to all TLD readers.

While the prime purpose of the heating cycle is to heat the phosphor until the electrons are liberated from the dosimetry traps, it is often additionally used to fade rapidly any low-temperature traps by means of a low-temperature hold (pre-read). At high exposure levels it is often necessary to use a high-temperature hold (anneal) to ensure complete erasure of all electrons from dosimetry traps. Figure 6.2 illustrates such a heating cycle used to read out LiF:PTFE disc dosemeters. The recorded integrated TL signal is denoted by the shaded area. PTFE disc dosemeters, because of their inherent flexibility, are particularly prone to the problem of variable thermal contact and in some reader systems a disc-retaining device is employed, such as a wire cruciform or spring.

The TL regenerative anneals such as those discussed in § 7.2 are generally performed in separate ovens. The heating cycles employed in different types of reader vary considerably; research readers should be capable of reading out many different phosphor materials with a wide range of glow peak temperatures, in a variety of physical forms with different thermal properties. Good glow curves are important, and they should be of high resolution suitable for analysis. The heater should therefore be capable of providing a wide range of linearly controlled heating and cooling rates; a choice of temperature plateaux; and flexibility in the choice of pre-read, read, anneal and cool periods.

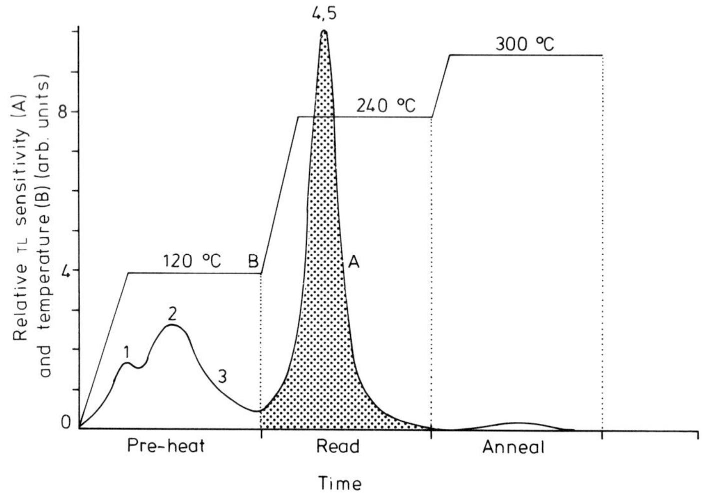

Figure 6.2 Heating cycle and glow curve for readout of LiF:PTFE disc dosemeters. Pre-heat temperature hold rapidly fades TL from the low-temperature peaks. TL integration is denoted by the shaded area. (Courtesy D A Pitman Ltd.)

On the other hand, high-throughput personal dosemeter reader systems require an acceptable precision, reproducibility of integrated signal and little or no glow curve data. The heating must therefore be fast, reproducible and not necessarily linear. Most commercial readers offer a compromise between these two extremes. Some have heating cycles with fixed parameters, but offer flexible research heating modules as plug-in accessories.

Let us now briefly examine the various types of heating used.

6.2.1 Ohmic heating

Ohmic heating, as the name implies, is simply the heating of a readout tray on which the dosemeter is placed, either directly by the passage of a current through the tray, or indirectly by bringing it into contact with an electrically heated element (low thermal capacity heating) or block (high thermal capacity heating). In the direct case the heating tray is built into the reader assembly and forms a part of the secondary winding circuit of a step-down transformer. The tray can generally only be removed by the use of an appropriate tool. Regulation of temperature is achieved by means of a thermocouple in contact with or preferably welded to the underside of the tray. In the indirect case the heating tray containing the dosemeter is placed in an aperture in a drawer assembly which, when pushed in, brings it into contact with a heating element or preheated block. Temperature regulation is again achieved by means of a thermocouple.

Ohmic heating is by far the most commonly used form of heating.

6.2.2 Hot nitrogen gas

A number of readers have been built utilising heated nitrogen gas as the heat transfer medium (Böttor-Jensen 1970, Julius *et al* 1974). The heating is very rapid and has proved particularly suited to automated readers, although it is only suitable for solid form dosemeters and not for loose powder.

6.2.3 Radio-frequency heating

With radio-frequency heating (Brunskill 1968), the TLD is placed on graphite or other suitable material which is heated by the induced current produced by an RF induction heating coil. Although the method has proved reliable it has not found widespread application probably because of the difficulty in variably controlling the heating cycle. Direct temperature measurement is not possible.

6.2.4 Radiant heating

The radiant energy from an infrared heating element or focused projection bulb can be used to heat the tray. Cameron *et al* (1968) described a relatively simple reader incorporating such a heating system using a 150 W projection bulb. A major drawback of such a system is the requirement to eliminate leakage of light into the detector system, and the method has not found widespread application. Recently, however, the Matsushita Co. Ltd have developed a personal dosemeter reader using a pulsed infrared source capable of very fast heating, as illustrated in figure 6.3. The dosemeter is in the form of phosphor granules deposited onto a carbon-coated polyimide film.

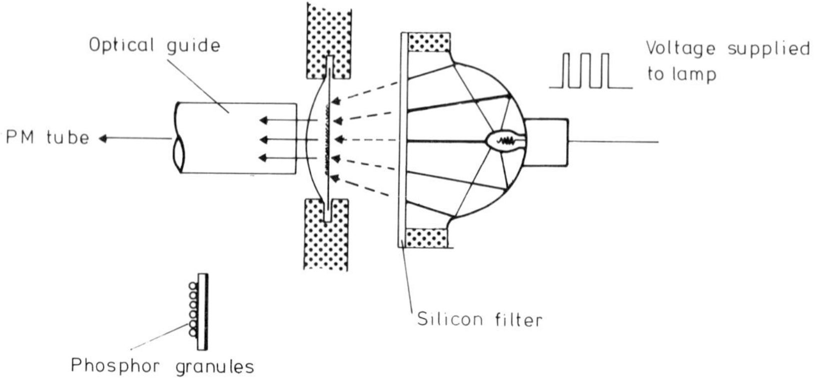

Figure 6.3 Radiant heating system. The voltage is supplied to the lamp as a series of 10 V pulses of 0.37 s duration. The 1 mg sample of $CaSO_4$:Tm powder can be read out in less than 1 s. (Reprinted courtesy of Matsushita Electric Industrial Co. Ltd.)

6.3 Light Collection, Detection, etc

The purpose of the light collection and detection system is to collect the thermoluminescence emitted by the phosphor as efficiently as possible rejecting all other optical radiations such as infrared thermal glow from the heating tray, and converting it into an electrical signal (e.g. current), integrated charge or voltage pulses, suitable for display and recording. The peak luminous flux of the thermoluminescence emission from a dosemeter corresponding to an absorbed dose of a few mGy is typically 10^{-12}–10^{-10} lm†. The only suitable light detector which can detect and convert such small quantities of light and which has a sufficiently large sensitivity range enabling a uniform response over the luminous flux equivalent of six decades of absorbed dose, is the photomultiplier.

As we have seen in Chapter 3, the spectral emissions of TL phosphors differ considerably but for all commonly used dosimetry materials they fall within the range of wavelengths 270–700 nm. The spectral response of a photomultiplier depends on the composition of the photocathode and on the spectral transmission of the tube window. The spectral responses of several types of photocathode are shown in figure 6.4. All of the tubes have a peak sensitivity around 400 nm, corresponding with

† The use of the photometric term lumen (lm) in this context is not entirely appropriate as it refers to the photopic response of the human eye (peak 550 nm). However, it is used by PM tube manufacturers.

the blue emission of LiF. For the detection of the red 600 nm emission of $Li_2B_4O_7$:Mn phosphor a tube with an S-20 or extended S-20 photocathode would provide greater sensitivity, but would also produce an increase in unwanted thermal signal from the heating tray unless additional spectral discrimination was used. By using the more expensive

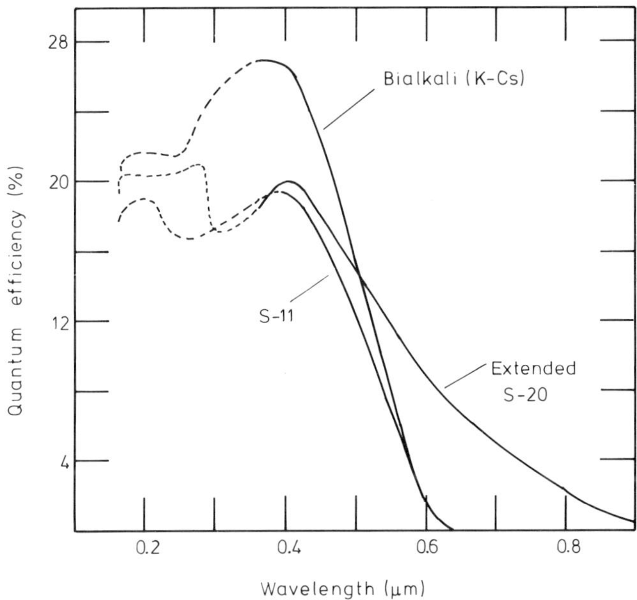

Figure 6.4 Relative spectral sensitivities of three types of photomultiplier tubes. Broken lines denote fused-silica window tubes. (EMI 1979, produced with the permission of EMI Ltd.)

quartz (fused-silica) window material instead of the standard soda or borosilicate glass, the response can be extended well into the ultraviolet. To provide spectral discrimination between the phosphor emission and unwanted thermal radiation a cut-off or band pass filter is often used. The majority of photomultiplier tubes are either 1 or 2 inch diameter tubes with either side or end windows. Most tubes used in TLD readers are 11 or 13 stage 2 inch diameter end window tubes which are chosen for three main reasons: (i) the large solid angle subtended by the 2 inch tube cathode ensures high light collection efficiency; (ii) close coupling and symmetrical geometrical efficiency is obtained, and (iii) the gain is high—typically 10^6.

The gain of a photomultiplier tube is critically dependent on the number of dynode stages and on the voltage applied between successive pairs of dynodes. The relationship between the gain stability and the voltage stability for a tube with N stages is given by

$$\left(\frac{\Delta G}{G}\right) = 0.7\,N\left(\frac{\Delta V}{V}\right).$$

Typically, for an 11 or 13 stage tube the fractional change in gain will be approximately ten times the fractional change in voltage. To maintain a $\pm 1\%$ stability in gain therefore requires a voltage supply stabilised to $\pm 0.1\%$. The gain stability is dependent on the dynode configuration of the tube and it has been found that the so-called venetian blind dynode system gives the optimum long-term gain stability (EMI 1979).

When a photomultiplier is operated in complete darkness some electrons are emitted from the photocathode mainly by thermionic emission. This phenomenon sets a limit on the lowest detectable intensity of TL emission. The current due to these electrons is called the dark current. It may be reduced by cooling the tube below normal ambient temperature. The problem then arises of maintaining the lower temperature within close limits as the gain of the tube is a function also of temperature; for a bialkali photocathode it is typically $1\%\ K^{-1}$. For this reason the photomultiplier is often mounted in the reader with its window plane normal to the plane of the heating tray and heat filter, thus avoiding air convection heating of the tube. The TL emission is reflected by a $45°$ mounted mirror. Similarly, light coupling pipes have been used.

A convenient method of tube cooling often used in commercial TLD readers is a thermoelectric cooler with heat sink. A particular problem sometimes encountered in the use of cooled tubes is the condensation of water vapour onto the photomultiplier base. This can result in high-voltage tracking but can often be overcome by placing a sachet of hygroscopic silica gel in the tube housing. The operation of a photomultiplier involves the passage of electrons along the length of the tube. As charged particles they are influenced by the presence of external electric and/or magnetic fields. The effect can be minimised by the use of a Mu metal shield surrounding the photomultiplier and maintained at cathode potential. If a photomultiplier is subjected to prolonged or repeated exposure to a source of light equivalent to a tube current of $>10^{-6}$ A, fatigue may result. This leads to a gradual loss of gain and an increase in dark current, hence reducing the signal to noise ratio of the system.

6.4 Current Measurement

The thermoluminescence can be measured and displayed in several ways, as illustrated in figure 6.5.

In the pulse counting mode the photomultiplier current which for any dosemeter is proportional to the thermoluminescence luminous flux, is converted into a series of fixed-amplitude voltage pulses whose pulse repetition frequency is uniformly proportional to the current over many decades (see for example Perry 1968). The frequency of the pulse train can be measured by a ratemeter which provides a proportional 10–100 mV DC voltage signal to display as TL on a chart recorder. The total

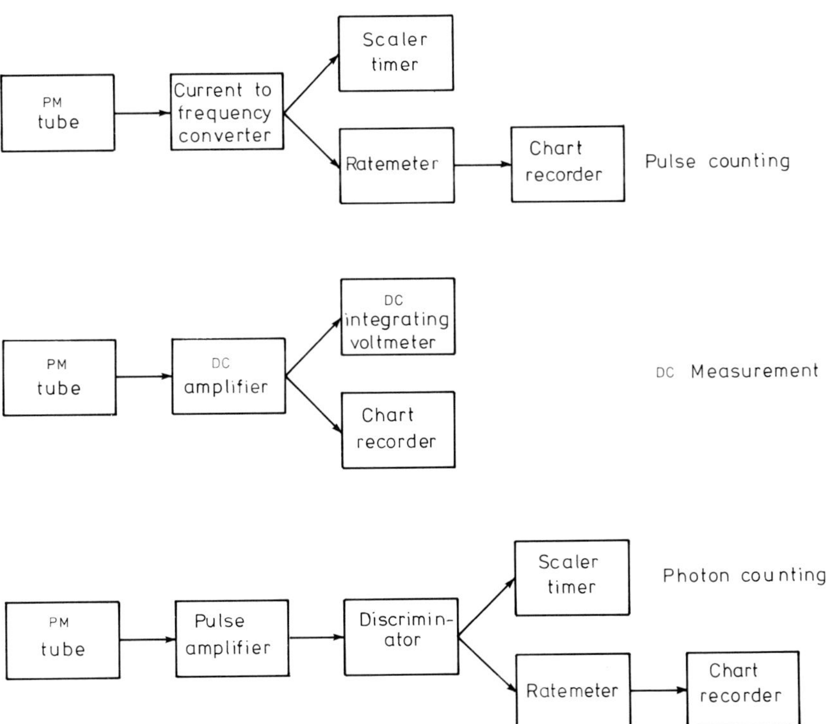

Figure 6.5 TL signal measurement techniques.

number of pulses corresponding to the integrated thermoluminescence emitted during the readout cycle can be counted and recorded on a pulse scaler.

In the current measuring/charge integration mode the photomultiplier current is amplified by a DC amplifier (electrometer) and the amplified current is displayed as thermoluminescence intensity on the chart recorder. Integration of the thermoluminescence is achieved by a DC integrating voltmeter which effectively collects all the charge on a series of capacitors.

The third method illustrated in figure 6.5 is photon counting. Each pulse formed at the photomultiplier anode corresponds to a single photoelectric event at the photocathode. If the gain of the photomultiplier varies, only the amplitude of the pulse changes, and not the frequency. Pulses of a range of heights known to correspond to thermoluminescence-produced pulses are selected by a discriminator which rejects low-amplitude pulses associated with thermionically generated dark current. The reduction in dark current achieved by this method is not great, however (Spanne 1974). The photon counting technique has found application in thermoluminescence dating but is not generally used in commercial dosimetry readers.

6.5 Commercial Systems

To review all the currently available commercial TLD readers would clearly be outside the scope of this text and would be quickly dated by design changes and performance modifications. Only a broad outline of commercial systems will be presented with a little detail where appropriate. Comparisons of reader performance are deliberately excluded. Detailed performance specifications can be obtained from the manufacturers listed in Appendix I.

6.5.1 Multi-dosemeter readers

These readers are capable of reading a variety of different phosphors in various physical forms. Some are manually operated, others are semi- or fully automated. Because of their inherent adaptability they have found widespread use in numerous fields, including personal monitoring, clinical dosimetry and environmental monitoring.

Examples of manual readers are the Harshaw 2000, Pitman 654, Teledyne 7300 and the Victoreen 2810. All of these are bench-top readers (see figure 6.6). They all employ ohmic heating with varying degrees of selection of heating cycle parameters. While they are different

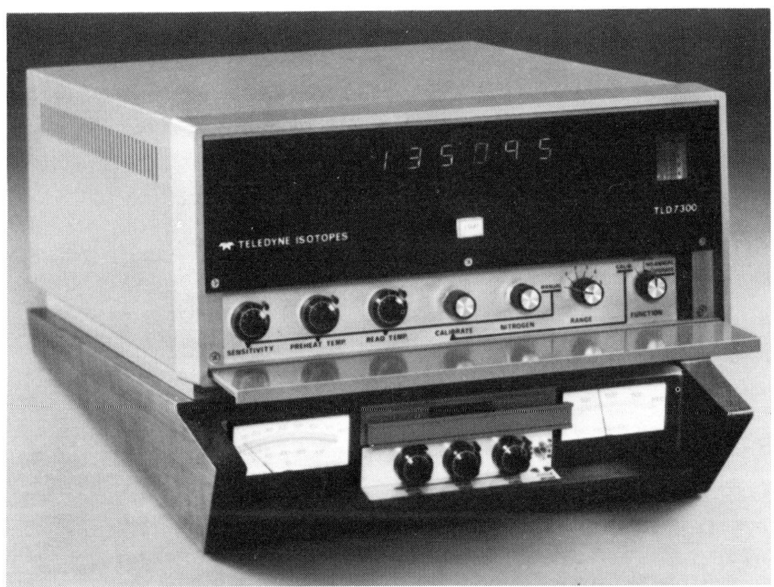

Figure 6.6 An example of a bench-top manual TLD reader (Teledyne Isotopes TLD 7300). It can be used to read out many forms of phosphor, including powder, chips, discs, micro-rods, etc. (Photograph by courtesy of Teledyne Isotopes Inc.)

with regard to specific details of design and performance, they all have the following features in common:

(i) they can read out a variety of forms of dosemeter,

(ii) the absorbed dose measurement range extends from 10^{-6}–10^{4} Gy,

(iii) they provide BCD printer output for integrated thermoluminescence recording, and

(iv) they offer a wide variety of accessories, options and modifications.

The Harshaw Atlas reader is an example of an automated multi-dosemeter system (see figure 6.7). It is capable of handling up to 50 individual rod, ribbon and chip dosemeters, indexed and read at the rate of four per minute. The heating is by hot nitrogen gas.

Figure 6.7 The Harshaw Atlas automated TL reader. Up to 50 TL elements can be read out from a single loading. They are indexed and read at a rate of four per minute. (Photograph by courtesy of Harshaw Chemical Co. Inc.)

6.5.2 Single dosemeter designated systems

These readers are invariably high-throughput semi- or fully automated systems. They are capable of reading only dosemeters of one specific design and are used mainly but not exclusively in the field of personal dosimetry.

The dosemeters vary in form but are usually plates or cards containing two or more individual thermoluminescent elements. For personal monitoring the plate or card is contained inside a badge holder.

The thermoluminescent elements are of various types, e.g. the Harshaw 2271 system uses a dosemeter card containing two chips, which may be of a variety of phosphors, sandwiched between two films of PTFE. The Pitman 605 reader (figure 6.8) uses an NRPB-designed plate dosemeter containing either one 0.2 mm thick and one 0.4 mm thick LiF:Mg:Ti:PTFE discs for personal monitoring, or two 0.4 mm thick $CaSO_4$:Dy:PTFE discs for environmental monitoring. The Teledyne Isotopes 9100 reader uses a single piece of PTFE-based phosphor with four separate readout areas and four further contingency read areas. Again the dosemeter is worn in a badge.

Figure 6.8 The Pitman 605 automated TL reader. The operator is loading the dosemeter inserts (similar to that shown in figure 6.9) into a magazine prior to readout. Each magazine holds a maximum of 200 inserts, and the maximum throughput, including pre-heat and partial high-temperature anneal, is 280 per eight-hour working day. (Photograph by courtesy of D A Pitman Ltd.)

6.5.3 Purpose-built automated systems

A number of such systems have been built and are operated by organisations requiring a high-throughput personal dosimetry system linked to a computer. The capital cost of the design and manufacture of such systems is very high but can be offset by the savings in manual labour, increased efficiency and precision in data handling.

The National Radiological Protection Board (NRPB) operate such a system in the UK (Dennis *et al* 1974). The dosemeter used consists of an aluminium alloy plate insert containing two LiF:Mg:Ti:PTFE discs, as shown in figure 6.9. The thermoluminescence sensitivity of each disc is regularly checked and recorded. To obviate the problems of light-induced thermoluminescence associated with PTFE-based dosemeters, each dosemeter is sealed in a black plastic bag prior to issue. The wearer of the dosemeter inserts it into a plastic badge holder which he retains. This is also shown in figure 6.9.

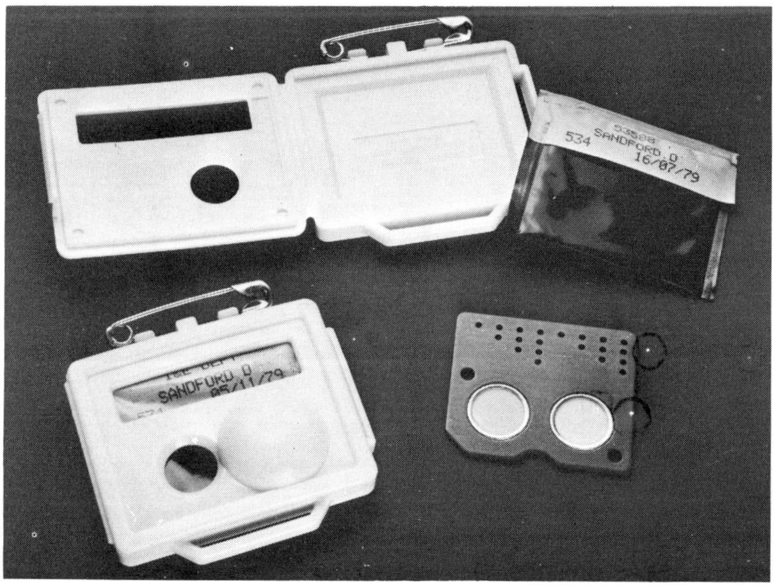

Figure 6.9 The NRPB dosemeter insert and badge holder. The 0.2 mm (55 mg cm^{-2}) disc on the left of the insert lies under the open window of the holder. The 0.4 mm (110 mg cm^{-2}) disc lies under the 700 mg cm^{-2} domed plastic filter. (Photograph by courtesy of the National Radiological Protection Board, Harwell.)

The manufacture of coded inserts, dosemeter calibration, packing, issue, contamination check, unpacking, readout, disc optical opacity check, 300 °C anneal, dose calculation, dose recording and glow curve display, updating of the appropriate dose records and issue of dose reports, etc, to the customer is controlled by computer. The readout system is shown in figure 6.10.

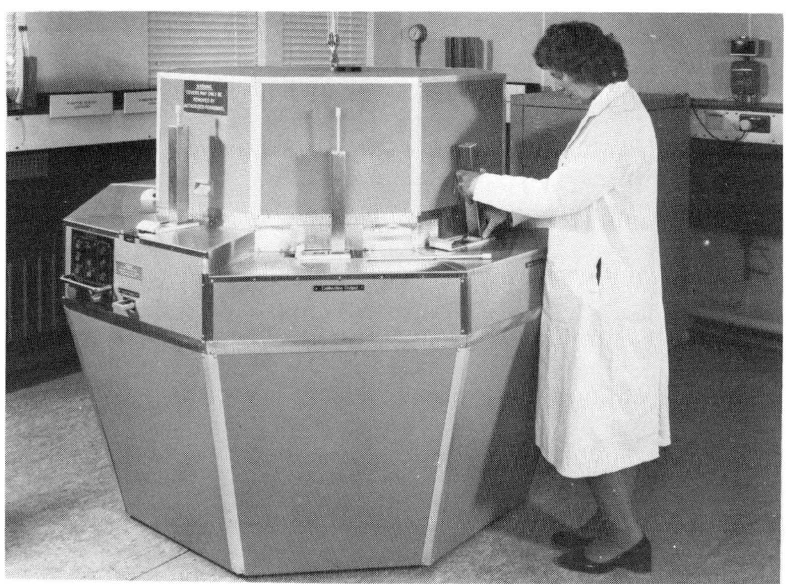

Figure 6.10 The NRPB automated readout system. The dosemeter inserts (see figure 6.9) are loaded into magazines. The input stage is on the left. The inserts then rotate clockwise around the reader, undergoing an automatic sequence of (i) readout of both elements, (ii) optical density check, (iii) 15 s anneal. The operator is removing inserts from the output stage. (Photograph by courtesy of the National Radiological Protection Board, Harwell.)

7 Experimental Problems and Handling Techniques

7.1 Introduction

While many potential sources of variations in the thermoluminescence characteristics of dosemeters are due to their manufacture and are outside the control of the user, there are many others which are clearly within his control. A complete cycle of use of a TL dosemeter consists of a combination of

(1) annealing,
(2) storage and handling,
(3) irradiation,
(4) readout.

There are opportunities for making mistakes at all of these stages. It is the purpose of this chapter to identify those areas where errors or less than careful experimental techniques can lead to low precision and non-reproducibility of results.

7.2 Annealing of Dosemeters

All phosphors display some changes in their thermoluminescence characteristics depending on the thermal treatment which they receive. To ensure complete readout of stored signal and repeated use of the phosphor without significant change in its thermoluminescence sensitivity, a thermal anneal is almost always required. Before making radiation measurements, therefore, all dosemeters should be identically annealed, as far as is practically possible, to standardise their sensitivities and backgrounds. For some phosphors the anneal may be simple—such as for $Li_2B_4O_7$:Mn—but for others it may be complex, such as for LiF:Mg:Ti. In LiF:Mg:Ti pre-irradiation annealing is especially important in order to remove all residual TL signal, to establish the TL sensitivity, and to eliminate unstable low-temperature glow peaks. A comprehensive study of the annealing characteristics of TLD 100 by Zimmerman *et al* (1965) confirmed the optimum anneal as 1 h at 400 °C, followed by 16–24 h at 80 °C. The effects of temperature variations within the range 80–400 °C are shown in figure 7.1. It has also been observed that repeated 1 h 400 °C anneals produce a decrease in TL sensitivity in LiF extruded-ribbon dosemeters of up to 18% after 100 cycles (Wald *et al* 1977). The effect of the 80 °C anneal is particularly important with regard to the elimination of the low-temperature glow peaks.

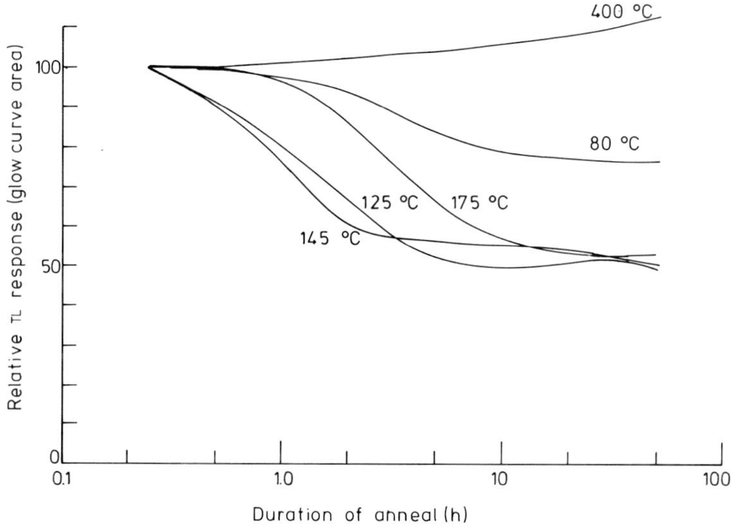

Figure 7.1 Effect of various pre-irradiation annealing regimes on the TL sensitivity of LiF:Mg:Ti (Zimmerman *et al* 1965).

While this anneal regime is an optimum, the circumstances of the use of dosemeters may preclude its use; for example, if the phosphor is held in a PTFE matrix it cannot be annealed at temperatures much above 300 °C. If dosemeters are to be re-used quickly, a 16h anneal at 80 °C may not be possible. Alternatively, a 1 h 100 °C anneal may be used (Mason *et al* 1976). If the low-temperature anneal is omitted entirely, the relative magnitude of the low-temperature peaks will depend critically on the cooling rate from the high-temperature anneal, as shown in figure 2.6. Regulla (1971) has proposed a low-temperature post-irradiation anneal of 10 min at 100 °C to fade rapidly the low-temperature traps. For this method to produce consistent results very precise temperature control is required. Alternatively, a suitable reader pre-read such as that described in § 6.2 may be used.

For PTFE-based dosemeters this technique also reduces the effects of spurious phosphorescence and thermoluminescence induced by ambient ultraviolet radiation and visible light. Where relatively low absorbed doses are measured, a long high-temperature anneal may not be necessary as any residual TL, which may account for 5–10% of the measured signal, can be eliminated by a short high-temperature post-read hold in the readout cycle (Marshall *et al* 1971); for example, 16 s at 300 °C should typically reduce the residual signal to < 0.1%. For dosemeters which have received a very high absorbed dose, perhaps as a result of radiotherapy exposure, a high-temperature anneal should be used. At absorbed dose levels where a significant degree of sensitisation may occur a high-temperature anneal is necessary to remove all the TL signal from the deep traps. This presents a problem with PTFE-based dosemeters. After an absorbed dose of 10 Gy, a 16 h anneal at 300 °C is insufficient to remove sensitisation (Linsley and Mason 1971).

Although individual dosemeters may be annealed in the reader, when a long-term anneal is required or many dosemeters have to be annealed, an external annealing oven is used. Most TLD equipment manufacturers supply suitable ovens as approved accessories to their readout systems. Some of these ovens are of the fan-assisted hot-air circulation type. This heating method ensures a rapidly achieved uniformity of temperature throughout the entire volume of the oven, and hence a reduction of temperature gradients in the dosemeters. Ovens should be kept scrupulously clean, and preferably should be used for only one type of phosphor to prevent cross contamination and intermixing.

Dosemeters should be annealed in suitable containers, a wide range of which are in use. Figure 7.2 shows three types. The disc stack type is commonly used to anneal PTFE-based disc dosemeters in batches of more than 200 up to a temperature of 300 °C. In order to retain their flatness the dosemeters are sandwiched between aluminium plates as shown. The stainless steel block is suitable for extruded-ribbon form dosemeters. Steel, nickel, glass or porcelain (shown in the photograph) crucibles are used for the bulk annealing of powder, extruded dosemeters and PTFE-based micro-rods. Variations in the temperature of the oven may often be overshadowed by the effect of the thermal inertia of the annealing container (illustrated in figure 7.3). It is important therefore

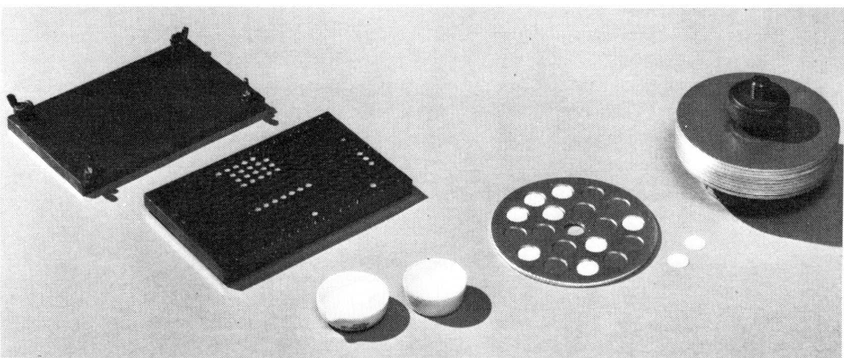

Figure 7.2 Containers used for annealing of TL dosemeters. (Photograph by courtesy of the National Radiological Protection Board, Harwell.)

to add sufficient time to the total anneal time to allow dosemeters to reach their chosen anneal temperature. The cooling rates of dosemeters allowed to cool in the different types of annealing containers are also variable. Phosphors annealed in open crucibles attain their anneal temperature much more quickly and cool rapidly and reproducibly when removed from the oven. It has been shown that a series of at least four initialisation cycles comprising readout, a 16 s 300°C in reader anneal,

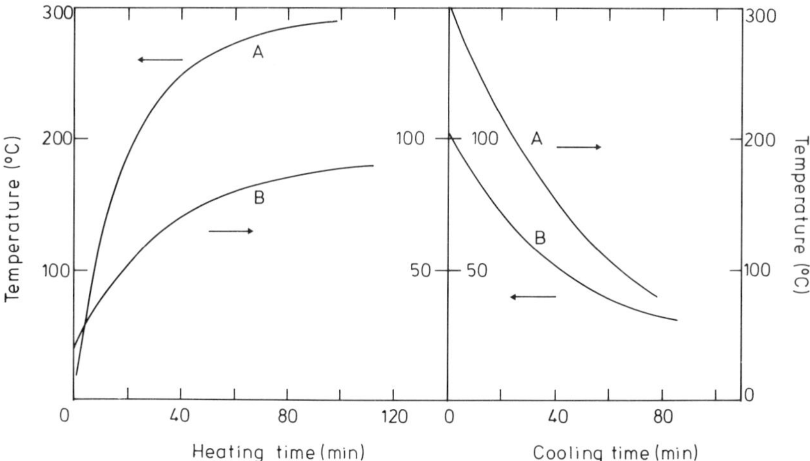

Figure 7.3 Heating and cooling curves for PTFE disc dosemeters annealed at A, 300°C, and B, 100°C in a stack.

and a 16 h 80°C oven anneal stabilises the sensitivity of LiF extruded-ribbon and LiF : PTFE dosemeters (Bartlett 1979).

7.3 Storage and Handling

Many aspects of the storage and handling of dosemeters can affect their TL sensitivity, stability, precision and minimum detectable absorbed dose. The major effects produced by storage and handling of dosemeters can be broadly divided into those due to

(1) environmental factors such as temperature, humidity, ultraviolet and visible radiation, and other agents;

(2) physical handling factors such as sieving, dispensing, picking up, cleaning, sterilising, etc.

There are obviously some factors which overlap these divisions.

7.3.1 Environmental factors

The thermal fading characteristics of phosphors are discussed in detail in Chapter 3. Care should be taken to ensure that in storage and use,

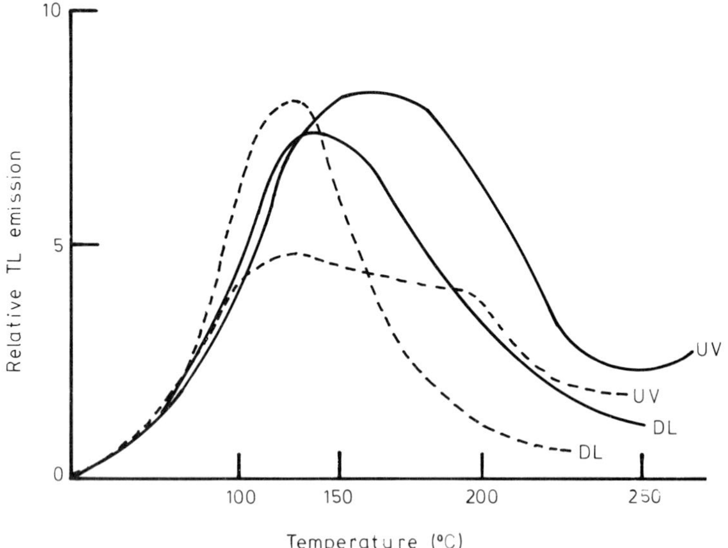

Figure 7.4 Glow curve structure of UV- and daylight (DL)-induced TL in LiF:PTFE discs (full curves), and pure PTFE discs (broken curves) (Robertson 1975).

dosemeters are not heated much above normal ambient temperature. Dosemeters which are particularly affected by humidity, such as those using $Li_2B_4O_7$: Mn phosphor, should be stored with a desiccating agent when not in use, and sealed in suitable containers for use, such as the polythene sachets shown in figure 4.12.

Many phosphors respond to normal ambient levels of ultraviolet and visible radiation. The effects are twofold: the production of a light-induced TL signal, and the phototransfer and subsequent retrapping of trapped charge carriers. In some phosphors the latter effect can result in increased fading of the dosimetry traps, while in others a transfer of electrons to the dosimetry traps results in an apparent increase in the subsequently recorded TL signal. PTFE is particularly sensitive to long-wavelength ultraviolet radiation (UV 315–400 nm), resulting in induced phosphorescence and thermoluminescence signals. TL glow curves obtained from pure and LiF-loaded PTFE discs exposed to ultraviolet radiation (365 nm) and visible light, are shown in figure 7.4.

The packaging of PTFE-based dosemeters in light-tight envelopes, such as those illustrated in figure 4.12 obviates these effects. Ambient lighting levels, especially from fluorescent lights, should be reduced in areas where PTFE dosemeters are handled or processed outside their protective envelopes.

7.3.2 Physical handling factors

Some of the properties of different forms of dosemeter are discussed in § 3.7. For the purpose of discussing possible uncertainties in precision and reproducibility associated with the handling of dosemeters, it is assumed that the choice of a particular form of dosemeter is based primarily on dosimetric considerations.

Within limits the TL sensitivity of a dosemeter is directly proportional to the mass of active phosphor present. While for solid form dosemeters, such as extruded ribbons, rods and PTFE-based discs, etc, the mass of the active phosphor present in each dosemeter is fixed during manufacture, and if carefully handled should not change; for powder dosemeters the dispensing of the chosen mass of powder is in the hands of the user. Ideally each sample of phosphor should be weighed immediately prior to readout while positioned on the readout tray, and in this way the TL readout may be corrected for weight variations. Although this procedure is often adopted for research purposes when high-precision measurements are required, it would be extremely time-consuming to adopt for routine measurement of many dosemeters. For this a volume dispensing method is often used. Many different types of powder dispenser are commercially available. In general a reproducibility of better than ± 1% is claimed for most powder dispensers, but care should be taken to check their precision periodically, since changes can occur due to normal wear and tear. Fine powder is generally more difficult to dispense because the fine grains tend to adhere to the dispenser and to each other. For LiF and $Li_2B_4O_7$ phosphors a range of grain sizes 75–200 μm is generally used; these grains run quite freely.

Dispensers should be kept clean, used for one type of phosphor only (or cleaned and dried thoroughly before use with another), and stored with a desiccating agent when not in use. The selection of a reasonably narrow range of grain sizes is also important because of the observed dependence of TL sensitivity on grain size below 100 μm (King 1974, Yamaoka 1978).

While the handling of extruded-ribbon and PTFE-based dosemeters appears simple, care should be taken not to scratch or abrade their surfaces. Although extruded ribbons are fairly rugged they should not be dropped onto a hard surface as they may chip or fracture. The effect of the repeated use of steel forceps for handling extruded ribbons has been studied; Cox *et al* (1976) observed that a roughly handled extruded-ribbon dosemeter could lose up to 25% of its TL sensitivity over 50 cycles of use, as illustrated in figure 7.5. The change in TL sensitivity could not be attributed to weight loss, which amounted to only 3%, but observation of the surfaces of the dosemeter under a microscope revealed the presence of many tiny scratches which produced a somewhat opaque appearance. These effects can be eliminated by using vacuum forceps.

While one should attempt to keep dosemeters clean, some dust or grease may become attached to them. It is important that this is removed before it is burned permanently into the surface. Loose dust can be removed by gentle air blowing or sucking. If the contamination is more firmly attached to the surface then gentle rinsing in pure grade methyl alcohol may be effective in removing it. If the dosemeters are badly stained then as a last resort they can be washed in chloroform. Chloroform can be a staining agent in its own right, and its staining action has been observed in LiF:PTFE disc dosemeters (Mason *et al* 1975). Such dosemeters should be rinsed thoroughly in methyl alcohol and distilled water. Spanne (1979) has demonstrated that washing in methanol containing 12 mol HCl/m^3 effectively reduces non-radiation-induced light emissions associated with surface contamination effects in lithium fluoride dosemeters. The use of ultrasonic cleaners is often recommended but the author has found their use somewhat limited. Increases in optical density, up to 5% losses in weight, and up to 25% losses in TL sensitivity, have been observed in a batch of LiF:PTFE dosemeters following ultrasonic cleaning (Mason *et al* 1975).

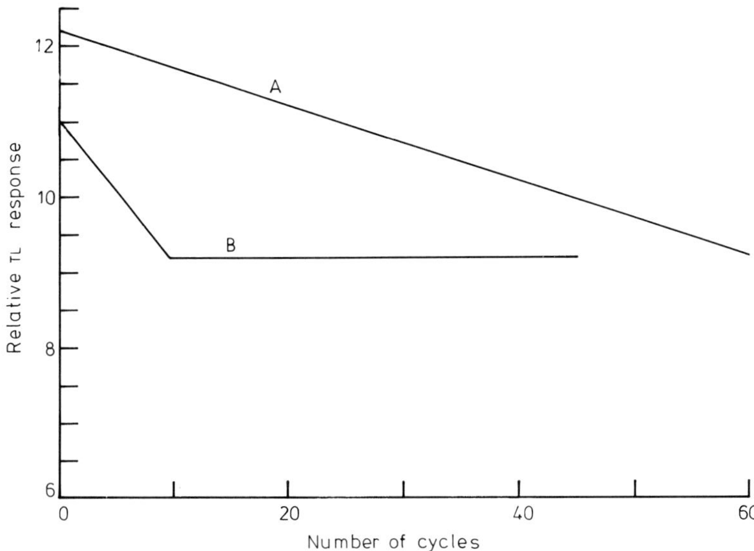

Figure 7.5 The effect of repeated handling on the TL sensitivity of LiF chips. A, recycling of chips using standard laboratory forceps; B, effect of changing to vacuum forceps after ten cycles. (Cox *et al* 1976, reprinted with the permission of Pergamon Press Ltd, Oxford.)

Dosemeters which are intended for clinical use must be packaged to protect them from a number of potentially adverse environmental conditions. Those intended for body surface measurements (either entrance or exit absorbed dose) may be heat-sealed in thin plastic sachets. The sachets can be taped directly onto the skin, with build-up as appropriate. The sachets protect the dosemeters from the effects of contact with skin moisture and grease, and the often highly photoluminescent adhesives used on tape. Micro-dosemeters intended for use for intra-cavitary measurement can be heat sealed in polythene catheter tubing, with radio-opaque markers if required. The tubing prevents attack by body fluids and possible toxic effects caused by dissolution of the phosphor. The dosemeters can also be either chemically or ultraviolet sterilised if required. Autoclaving should not be used as the combination of high temperature and humidity has a significant effect on the sensitivity of dosemeters.

7.4 Irradiation of Dosemeters

The reader is reminded that the concepts of exposure, absorbed dose, build-up, charged particle equilibrium, etc, which are used in this section are defined in Appendix II.

(i) *Calibration irradiation.* This is carried out to determine the TL response of the dosemeter to a measured exposure or absorbed dose of radiation of clearly defined energy. A ^{60}Co source is most often used for this purpose (gamma energies 1.17 and 1.33 MeV, half-life 5.3 years). As ^{60}Co decays at the rate of approximately 1% per month, a previously determined exposure rate will have to be corrected periodically.

(ii) *Routine laboratory irradiation.* This is done for a variety of reasons including the routine checking of dosemeter sensitivities following annealing, etc. Dosemeters can be graded and arranged according to their relative sensitivities; a set of dosemeters with closely matched sensitivities may be used for repeated checking of reader stability. The requirements for these types of irradiation are (1) reproducibility of absorbed dose, i.e. a fixed geometry and precise and reproducible exposure timing; (2) a fixed numerical relationship between a given laboratory irradiation and calibration absorbed dose; and (3) convenience of use. A ^{90}Sr–^{90}Y source (beta energies 0.61 and 2.2 MeV, half-life 28 years) is most suitable for this purpose. Some TL equipment manufacturers provide laboratory irradiation units fitted with exposure timers. Some of these units can be used to irradiate automatically a series of TL dosemeters for a pre-selected exposure time.

(iii) *Measurement irradiation.* This is the measurement of exposure or absorbed dose rate in a radiation field of type and energy not necessarily the same as that of the calibration field. However, the measurement irradiation conditions should match the calibration conditions as closely as possible.

For valid comparison of absorbed dose between any of the above irradiation conditions, all dosemeters must be read out under identical conditions.

7.4.1 Calibration of dosemeters with ^{60}Co radiation

TLDs are secondary dosemeters, i.e. they do not provide an absolute measure of radiation absorbed dose. Commonly used secondary dosemeters include photographic films, scintillation detectors, ionisation chambers and chemical dosemeters. All must be calibrated against a primary measurement system either directly or, more usually, via a calibrated secondary system.

TLDs are almost always calibrated against a calibrated ionisation chamber. Practical aspects of the calibration and use of ionisation chambers are described in a recent International Atomic Energy Agency publication (IAEA 1979).

The TL sensitivity calibration factor, C, for a TLD is given by

$$C = \frac{\mathrm{TL}}{(D_{\mathrm{w}})_{^{60}\mathrm{Co}}}, \qquad (7.1)$$

where TL is the measured TL signal from the dosemeter which has received an absorbed dose $(D_{\mathrm{w}})_{^{60}\mathrm{Co}}$ (Gy in water) of ^{60}Co photons. The response of a dosemeter at all other energies and for all other radiations may be expressed as a fraction or multiple of C.

C may be determined by exposure of the TLD to a beam of ^{60}Co photons in free air, as illustrated in figure 7.6(a). The exposure, $(X)_{^{60}\mathrm{Co}}$ is measured using a calibrated ionisation chamber, and the absorbed dose $(D_{\mathrm{w}})_{^{60}\mathrm{Co}}$ (Gy in water) is calculated from the expression

$$(D_{\mathrm{w}})_{^{60}\mathrm{Co}} = (X)_{^{60}\mathrm{Co}} \cdot F, \qquad (7.2)$$

where F is the Gy R^{-1} factor appropriate to ^{60}Co radiation. The dosemeter and the ionisation chamber must be surrounded by sufficient build-up material to ensure electronic equilibrium. Alternatively, for high-energy photons (including x-rays with generating potentials >150 kV), C can be obtained by exposure of TLDs together with an ionisation chamber in a water or water-equivalent phantom, as illustrated in figure 7.6(b). For ^{60}Co radiation, an irradiation depth of 5 cm is recommended. $(D_{\mathrm{w}})_{^{60}\mathrm{Co}}$ is calculated from equation (7.3). The value of $(X)_{^{60}\mathrm{Co}}$, the measured exposure, will differ from the free-air exposure as it will be increased by a contribution from scattered electrons from the phantom material, and reduced slightly due to absorption of the primary beam by the 5 cm phantom material.

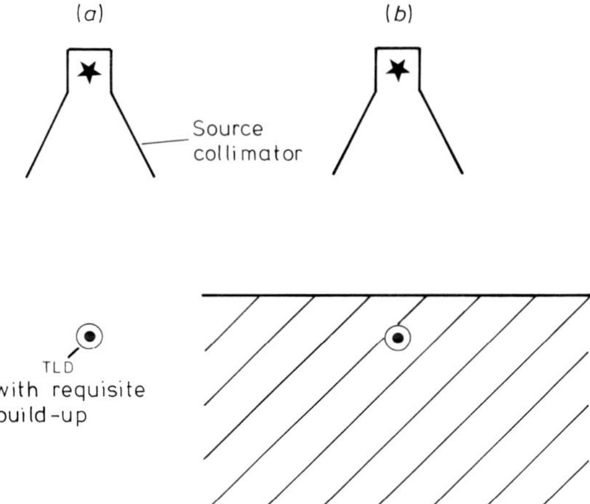

Figure 7.6 Exposure of TL dosemeters (*a*) in free air, and (*b*) in a water-equivalent phantom.

7.4.2 Calculation of absorbed dose

The measurement of absorbed dose in a material involves the introduction of a dosemeter into the material. Both the atomic composition and physical density of the dosemeter will generally be different from those of the material. There is therefore a discontinuity in the radiation absorption and scattering properties of the material at the material/dosemeter boundary. The dosemeter is said to represent a cavity within the material.

For a given exposure of radiation of energy E, the absorbed dose in the dosemeter $(D_{\mathrm{d}})_E$ is given by

$$(D_{\mathrm{d}})_E = (D_{\mathrm{m}})_E \cdot (S_{\mathrm{m}}^{\mathrm{d}})_E \tag{7.3}$$

where $(D_{\mathrm{m}})_E$ is the absorbed dose in the material, and $(S_{\mathrm{m}}^{\mathrm{d}})_E$ is the ratio of collision stopping powers for the dosemeter and material at energy E. The above expression is valid only for true Bragg–Gray cavities, i.e. where the cavity dimensions are small compared with the average range of the secondary electrons, and therefore does not perturb the electron spectrum in the material.

If the dosemeter cavity is large compared with the average range of secondary electrons, the absorbed dose in the cavity depends only on the absorption properties of the cavity, and the absorbed dose in the material depends only on the absorption properties of the material. Under these circumstances the absorbed dose in the dosemeter $(D_d)_E$ exposed to radiations of energy E is given by

$$(D_d)_E = \left[\frac{(\mu_{en}/\rho)_d}{(\mu_{en}/\rho)_m}\right]_E \cdot (D_m)_E, \tag{7.4}$$

where

$$\left[\frac{(\mu_{en}/\rho)_d}{(\mu_{en}/\rho)_m}\right]_E$$

is the ratio of the mass energy absorption coefficients of the dosemeter and the material at energy E.

The use of TLDs for absorbed dose measurements in the megavoltage photon and electron energy range involves an intermediate condition where neither equations (7.3) nor (7.4) apply, and where a generalised cavity theory must be used. Various generalised cavity models have been postulated in order to explain the observed apparent fall in TL sensitivity of dosemeters irradiated with megavoltage photons and electrons. In this case the absorbed dose $(D_d)_E$ in dosemeters exposed to radiation of energy E can be expressed as the absorbed dose in the material $(D_m)_E$ multiplied by a cavity correction factor, i.e.

$$(D_d)_E = (f)_E (D_m)_E, \tag{7.5}$$

where, for photons,

$$(f)_E = d\,(S_m^d)_E + (1-d)\left[\frac{(\mu_{en}/\rho)_d}{(\mu_{en}/\rho)_m}\right]_E \tag{7.6}$$

and d is a weighting factor for the relative contributions of primary radiation and secondary electrons generated inside the cavity. Numerically,

$$d = \frac{1-\exp(-\beta\bar{x})}{\beta\bar{x}} \tag{7.7}$$

where β is the energy absorption coefficient and $\bar{x}$ is the mean path of electrons through the cavity (Burlin 1966).

7.4.3 Calibration of dosemeters at other photon energies

In view of the difficulties of calculating the relative response of TLDs over a range of energies, including megavoltage photons and electrons, it is best, whenever possible, to calibrate the TLDs using radiation of the same energy as that required to be measured.

In principle, the TL sensitivity of a dosemeter can be measured using radiation of any energy, provided that

(1) the radiation exposure is measured using a calibrated ionisation chamber appropriate to the energy of the radiation;
(2) the appropriate value of the Gy (in water) R^{-1} factor is used (values of this factor for a range of photon energies are listed in table 7.1);
(3) build-up appropriate to the radiation is provided around the dosemeter and ionisation chamber to achieve electronic equilibrium.

Table 7.1 Values of Gy (in water) R^{-1} factor (ICRU 1976)†.

	Radiation quality	mGy (in water) R^{-1}
HVL	0.5 mm Al	8.9
HVL	1.0 mm Al	8.8
HVL	2.0 mm Al	8.7
HVL	6.0 mm Al	8.8
HVL	0.5 mm Cu	8.9
HVL	1.0 mm Cu	9.1
HVL	2.0 mm Cu	9.4
HVL	4.0 mm Cu	9.6
	^{60}Co	9.5
	2 MV	9.5
	6 MV	9.4
	12 MV	9.2
	18 MV	9.1
	30 MV	8.9

† Exposure in R (1 R $= 2.58 \times 10^{-4}$ C kg^{-1}).

The results of using insufficient build-up are twofold. In irradiations employing narrow beams of radiation the TLD will under-respond due to the non-attainment of electronic equilibrium. Over-response may occur due to absorption of electrons scattered primarily by the source housing and dosemeter holder, and to a much lesser extent by the air surrounding the dosemeter. This effect is greater for broad beams of radiation. Values of measured over-response factors for LiF extruded-ribbon dosemeters irradiated with photons of various energies are listed in table 7.2.

Table 7.2 Values of over-response factors for LiF chip dosemeters (3.2 × 3.2 × 0.89 mm). Source to TLD distance of 1.00 m (Graham and Homann 1978).

Radioisotope	Photon energy (MeV)	Over-response factor
^{137}Cs	0.66	1.02
^{226}Ra	Mean 0.8	1.06
^{60}Co	1.17 and 1.33	1.12
^{24}Na	1.37 and 2.75	1.08

7.4.4 Surface measurements

If dosemeters are calibrated for use in body surface absorbed dose measurements, they can either be calibrated against an ionisation chamber directly on the surface of a water-equivalent phantom, as illustrated in figure 7.7, or a factor appropriate to the photon energy can be applied to the free-air exposure in order to calculate the additional contribution from backscattered electrons. The surface absorbed dose, $(D_{w,s})_E$ (Gy in water) is given by

$$(D_{w,s})_E = (X)_E \cdot (F)_E \cdot (B)_E, \tag{7.8}$$

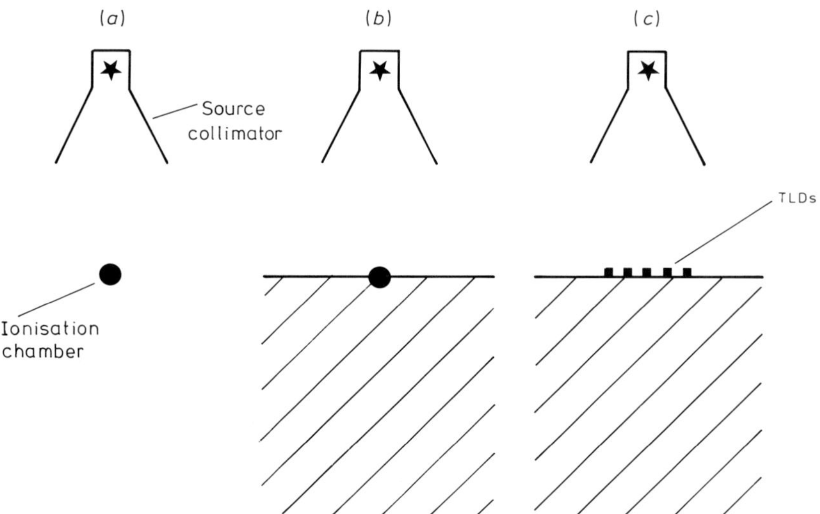

Figure 7.7 Exposure of TL dosemeters on the surface of a water-equivalent phantom.

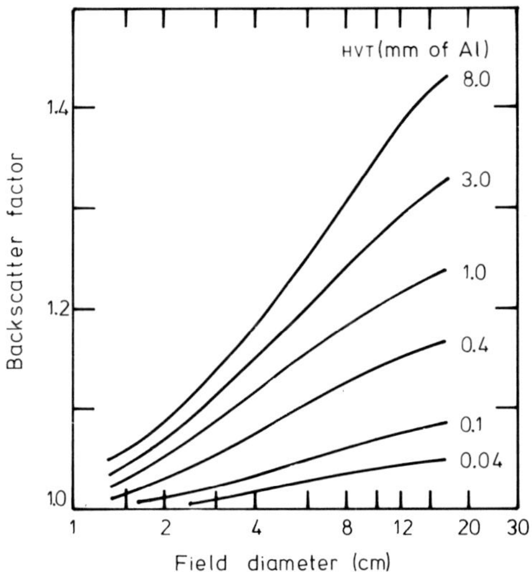

Figure 7.8 Variation of the backscatter factor with field diameter and half-value thickness in the range 0.04–8.0 mm Al. (Reprinted with the permission of the *British Journal of Radiology* 1978.)

where $(F)_E$ is the G R^{-1} factor, and $(B)_E$ is the backscatter factor at photon energy E. $(B)_E$ depends on the radiation energy and field size, as illustrated in figure 7.8.

7.4.5 Self-shielding

For radiation of photon energies below about 50 keV the effect of self-shielding in the dosemeter becomes increasingly important. It is essential that the orientation of dosemeters (whose three dimensions are not approximately equal) in the measurement field, should match as closely as possible their orientation in the calibration field. Alternatively, a measured correction factor can be applied. The measured TL response of $6.4 \times 6.4 \times 0.9$ mm LiF extruded-ribbon dosemeters, exposed with their square faces parallel and perpendicular to the axis of incident radiation, are shown in figure 7.9.

7.5 Readout of Dosemeters

As the basis of TL dosimetry is the comparison of TL light signals from dosemeters which have been exposed to an unknown absorbed dose of

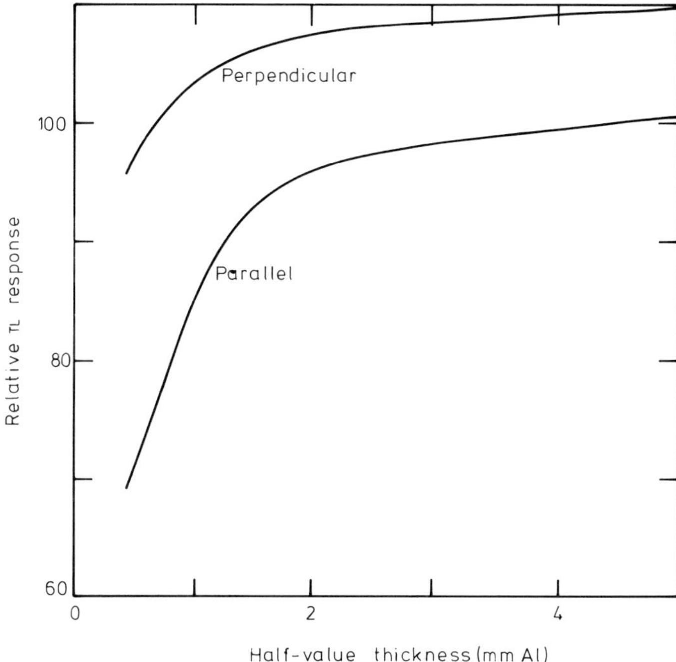

Figure 7.9 The measured TL response of LiF extruded-ribbon dosemeters exposed with their square faces parallel and perpendicular to the axis of the incident radiation. (Morgan and Bateman 1977, reprinted with the permission of Pergamon Press Ltd, Oxford.)

radiation, with those from similar dosemeters which have been given a calibration absorbed dose, it is important that the TL reader should be used in a reproducible manner.

The basic design characteristics of TLD readers are discussed in Chapter 6. Here we shall examine some of the potential sources of uncertainty related to their use, and detail what checks can be made to ensure that they are recognised or avoided.

7.5.1 Heating variations

One of the most common faults encountered in the readout of TLDs is bad and variable thermal contact between dosemeters and the heating tray. This usually manifests itself as a variation in the heating rate of the phosphor, or non-attainment of the chosen readout temperature, resulting in incomplete readouts. The problem is particularly important for PTFE dosemeters because they are flexible and (especially at fast heating rates) because they have low thermal conductivity. Bad thermal contact often results in a broadening of the glow curve with an increase in the high-temperature tail. Various designs of heating tray have been produced, each appropriate to a particular type of dosemeter, to minimise bad thermal contact.

7.5.2 Light detection variations

Stability of light detection is a prerequisite for readout precision and it is important that the user can recognise when variations or drifts in the light detection sensitivity have taken place. All commercial TLD readers have provision for checking the light detection sensitivity by means of a stable low-intensity light source. Some readers have internal light sources which are used to re-set automatically the sensitivity of the reader to a pre-set value between each readout (Robertson 1975).

Almost all light sources for use with TLD readers consist of a few microcuries of a long-lived radioisotope (often carbon-14, half-life 5570 years) mixed with a plastic scintillator. As the ageing characteristics of photomultipliers exhibit a wavelength dependence, it is important, where different TL phosphors with different TL emission spectra are being used, to choose a light source with an appropriate emission spectrum.

Because the intensity of emission of light sources may be affected by changes in temperature, ambient illumination levels, and the build-up of dust, etc, they should be kept clean, maintained at a constant temperature (20–25 °C), placed in opaque containers when not in use, and not exposed to high ambient light levels.

Light source checks should be routinely made before and after measurement sessions according to the manufacturer's recommendations. Any necessary adjustment can be made to the reader sensitivity, or a correction factor calculated and applied to subsequent readouts. However, further uncertainties can arise from

(1) variations in the reflectivity of readout trays (due to burnt surface dust, or oxidation);

(2) variations in the lateral positioning of dosemeters on the tray (avoided if appropriate depression is provided);

(3) thickness of dosemeter resulting in variations in TL self-absorption and scattering (this is especially important for the comparison of TL readouts obtained by irradiation with high LET radiation and with ^{60}Co and other low LET radiation).

7.5.3 Suppression of chemiluminescence and triboluminescence

The basic mechanisms involved in the production of chemiluminescence and triboluminescence in TL phosphors are obscure. An extensive review of triboluminescence effects has been written by Walton (1977). It is thought that chemiluminescence produced during readout may be due to oxidation of dirt which has been trapped on the surface of the dosemeter. The effect can certainly be reduced by clean handling. Triboluminescence is induced by virtually any form of mechanical disturbance of phosphor grains but is most noticeable after grinding and sieving of phosphor powder. Both effects are related to the surface and in general are much greater for dosemeters with high surface area to volume ratios such as powders and single crystals. The triboluminescence effects are minimal for extruded-ribbon form dosemeters but severe vibrations can cause them in PTFE-based dosemeters. The effects are shown as a series of extra non-radiation-induced glow peaks and the resultant uncertainties in background signal can severely limit the threshold of detection of dosemeters. Readout in an inert atmosphere almost entirely eliminates these effects, as illustrated in figure 7.10.

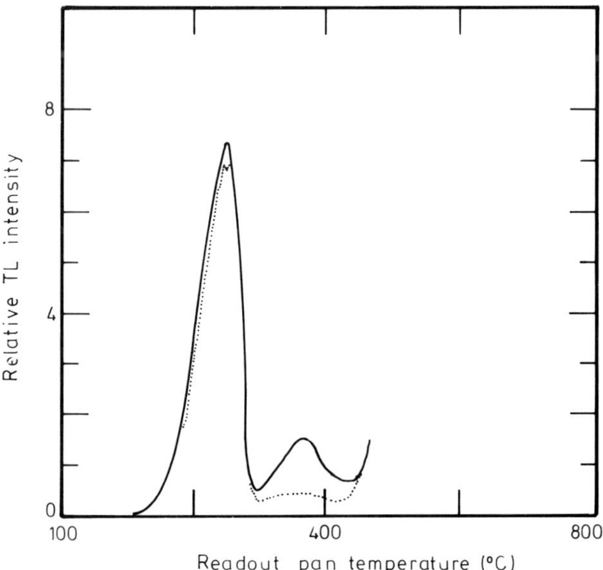

Figure 7.10 Glow curves for LiF (TLD 100) read out in air (full curve) and nitrogen (broken curve) (Nash *et al* 1965)

Nitrogen gas is cheaper than argon and is most often used. The gas should be free of oxygen and water vapour, and should be filtered between the storage cylinder and the TLD reader to remove any dust. The reader manufacturer will recommend a flow rate appropriate to the reader.

Table 7.3 TL results obtained from a set of 48 standard LiF extruded-ribbon dosemeters (TLD 700) subject to a series of three routine readouts. Absorbed dose 2 mGy ^{137}Cs gamma photons (0.67 MeV). All dosemeters were given four initialisation cycles comprising: readout, a 16 s 300 °C reader anneal, and a 16 h 80 °C oven anneal prior to use. A 1 h 400 °C and 16 h 80 °C anneal were then given prior to each irradiation.

Readout cycle	Absorbed dose			
	2 mGy		Zero dose†	
	Mean TL counts	% standard deviation	Mean TL counts	% standard deviation
1	404.4	2.95	8.0	23.5
2	406.7	2.65	10.0	36.7
3	410.0	3.39	12.0	24.7

† The calculated minimum detectable absorbed dose is 28 μGy, based on an uncertainty of two standard deviations in the TL from unirradiated dosemeters.

A series of checks on the efficiency of the entire readout system can be made using a set of 'identically' annealed and irradiated extruded-ribbon dosemeters. Dosemeters from the set can be read out before, after and at intervals during the readout session. Table 7.3 shows the results of three consecutive sets of readout measurements made with such dosemeters.

Appendix I Manufacturers of TL Phosphors, TLD Readers and Ancillary Equipment

BDH Chemicals Ltd, Poole BH12 4NN, UK

CEC, BP 60–92 Montrouge, France

Eberline Instrument Corporation, PO Box 2108, Santa Fe, New Mexico 87501, USA

Harshaw Chemical Co., 1945 East 97th Street, Cleveland, Ohio 44106, USA

Matsushita Electric Industrial Co. Ltd (National), PO Box 288, Osaka, Japan

D A Pitman Ltd, Mill Works, Jessamy Road, Weybridge, Surrey KT13 3LE, UK

Teledyne Isotopes Inc., 80 Van Buren Avenue, Westwood, New Jersey 07675, USA

Victoreen Instrument Division, 10101 Woodland Avenue, Cleveland, Ohio, USA

Appendix II
Terms and Definitions

Electron and charged particle equilibrium

Indirectly ionising radiations (e.g. photons and neutrons) produce a direct contribution to the absorbed dose in a material which is negligible compared with that from the secondary charged particles produced. If equal numbers of charged particles (either primary or secondary) of a specified energy are entering *and* leaving any small volume, then charged particle equilibrium exists within the volume. In the case where there are no primary charged particles then equal numbers of secondary charged particles of a specified energy may enter and leave the volume. This condition is called electronic equilibrium. In addition, if the number, type and energy of particles and/or photons entering a small volume in an absorber equals the number, type and energy of those leaving, then radiation equilibrium exists.

Practical achievement of charged particle equilibrium—'build-up'

To achieve charged particle equilibrium it is often necessary to employ 'build-up' material surrounding the volume to be irradiated. In all cases a sufficient thickness of build-up material should be used to ensure charged particle equilibrium without producing significant attenuation of the incident radiation. For all practical purposes no build-up is required for x-ray photons generated at tube potentials less than approximately 200 kV. However, the required thickness of build-up material increases with increasing voltage, as shown in table A1.

In relation to irradiation of the body, for x-rays with generating potentials less than approximately 200 kV the maximum delivered absorbed dose is to the surface of the body, i.e. the skin. For ^{60}Co gamma photons the maximum delivered absorbed dose is at a depth of approximately 5 mm.

Backscatter factor

The backscatter factor is the ratio of the absorbed dose rate on the surface of an object (e.g. a phantom) to the absorbed dose rate, at the same point in the beam, in free air. In reality, the absorbed dose rate is proportional to, and is calculated from the corresponding measured exposure rate.

Table A1 Required thickness of tissue-equivalent build-up material to ensure charged particle equilibrium.

Radiation	Thickness required (mm)
<200 kV	<1
^{137}Cs (0.67 MeV)	4
^{60}Co (1.25 MeV)	5
6 MV	15
22 MV	40

The following definitions and terms are those defined by the International Commission on Radiation Units and Measurements (ICRU 1980).

The *exposure*, X, is the quotient $\mathrm{d}Q/\mathrm{d}m$, where the value of $\mathrm{d}Q$ is the absolute value of the total charge of the ions of one sign produced in air when all the electrons (negatrons and positrons) liberated by photons in air of mass $\mathrm{d}m$ are completely stopped in air:

$$X = \mathrm{d}Q/\mathrm{d}m.$$

Unit: C kg^{-1}
The special unit of exposure, the roentgen (R), may be used temporarily:

$$1\ \mathrm{R} = 2.58 \times 10^{-4}\ \mathrm{C\ kg^{-1}}\ \text{(exactly)}.$$

The *absorbed dose*, D, is the quotient $\mathrm{d}\bar{\epsilon}/\mathrm{d}m$, where $\mathrm{d}\bar{\epsilon}$ is the mean energy imparted by ionising radiation to matter of mass $\mathrm{d}m$:

$$D = \mathrm{d}\bar{\epsilon}/\mathrm{d}m.$$

Unit: J kg^{-1}
The special name for the unit of absorbed dose is the gray (Gy):

$$1\ \mathrm{Gy} = 1\ \mathrm{J\ kg^{-1}}.$$

The special unit of absorbed dose, the rad, may be used temporarily:

$$1\ \mathrm{rad} = 10^{-2}\ \mathrm{J\ kg^{-1}}.$$

The *kerma*, K, is the quotient $\mathrm{d}E_{tr}/\mathrm{d}m$, where $\mathrm{d}E_{tr}$ is the sum of the initial kinetic energies of all the charged ionising particles liberated by uncharged ionising particles in a material of mass $\mathrm{d}m$:

$$K = \mathrm{d}E_{tr}/\mathrm{d}m.$$

Unit: J kg^{-1}
The special name for the unit of kerma is the gray (Gy):

$$1\ \mathrm{Gy} = 1\ \mathrm{J\ kg^{-1}}.$$

The special unit of kerma, the rad, may be used temporarily:

$$1\ \mathrm{rad} = 10^{-2}\ \mathrm{J\ kg^{-1}}.$$

The *mass energy transfer coefficient*, μ_{tr}/ρ, of a material for uncharged ionising particles is the quotient $(\mathrm{d}E_{tr}/EN)/\rho \mathrm{d}l$, where E is the energy of each particle (excluding rest energy), N is the number of particles, and $\mathrm{d}E_{tr}/EN$ is the fraction of incident particle energy that is transferred to kinetic energy of charged particles by interactions in traversing a distance $\mathrm{d}l$ in the material of density ρ:

$$\frac{\mu_{tr}}{\rho} = \frac{1}{\rho EN}\frac{\mathrm{d}E_{tr}}{\mathrm{d}l}.$$

Unit: m^2 kg^{-1}

The *mass energy absorption coefficient*, μ_{en}/ρ, of a material for uncharged ionising particles is the product of the mass energy transfer coefficient, μ_{tr}/ρ, and $(1 - g)$, where g is the fraction of the energy of secondary charged particles that is lost to bremsstrahlung in the material:

$$\frac{\mu_{en}}{\rho} = \frac{\mu_{tr}}{\rho}(1 - g).$$

Unit: m^2 kg^{-1}

The *total mass stopping power*, S/ρ, of a material for charged particles is the quotient $\mathrm{d}E/\rho \mathrm{d}l$, where $\mathrm{d}E$ is the energy lost by a charged particle in traversing a distance $\mathrm{d}l$ in the material of density ρ:

$$\frac{S}{\rho} = \frac{1}{\rho}\frac{\mathrm{d}E}{\mathrm{d}l}.$$

Unit: J m^2 kg^{-1}
E may be expressed in eV and hence S/ρ may be expressed in

eV m^2 kg^{-1}. S is the *total linear stopping power*. For energies at which nuclear interactions can be neglected, the total mass stopping power is

$$\frac{S}{\rho} = \frac{1}{\rho}\left(\frac{dE}{dl}\right)_{col} + \frac{1}{\rho}\left(\frac{dE}{dl}\right)_{rad},$$

where $(dE/dl)_{col} = S_{col}$ is the linear collision stopping power and $(dE/dl)_{rad} = S_{rad}$ is the linear radiative stopping power.

The *linear energy transfer* (LET) or *restricted linear collision stopping power*, L_Δ, of a material for charged particles is the quotient dE/dl, where dE is the energy lost by a charged particle in traversing a distance dl due to those collisions with electrons in which the energy loss is less than Δ:

$$L_\Delta = dE/dl.$$

Unit: J m^{-1}
E may be expressed in eV and hence L_Δ may be expressed in eV m^{-1}, or some convenient submultiple or multiple, such as keV μm^{-1}.

The *dose equivalent*, H, is the product of D, Q and N at the point of interest in tissue, where D is the absorbed dose, Q is the quality factor, and N is the product of all other modifying factors:

$$H = DQN.$$

The SI unit for both D and H is joule per kilogram. The special name for the unit of dose equivalent is the sievert (Sv):

$$1 \text{ Sv} = 1 \text{ J kg}^{-1}.$$

The special unit of dose equivalent, the rem, may be used temporarily:

$$1 \text{ rem} = 10^{-2} \text{ J kg}^{-1}.$$

The *dose-equivalent index*, H_I, at a point is the maximum dose equivalent within a 30 cm diameter sphere centred at this point and consisting of material equivalent to soft tissue with a density of 1 g cm^{-3}.

Unit: J kg^{-1}
The special name for the unit of dose-equivalent index is the sievert (Sv):

$$1 \text{ Sv} = 1 \text{ J kg}^{-1}.$$

The special unit of dose-equivalent index, the rem, may be used temporarily:

$$1 \text{ rem} = 10^{-2} \text{ J kg}^{-1}.$$

Appendix III
Toxicity of TL Phosphors

The possible toxic effects of TL materials are important from two standpoints.

(i) No person should suffer harm in the everyday handling of the material. If a material is so toxic that special safe handling techniques have to be employed, this should be regarded as a disadvantage of the material.

(ii) If a dosemeter is implanted in tissue it is important that the patient suffers no ill effects from either acute exposure to the material, or, in the event of non-recovery of a dosemeter, chronic exposure.

Ill effects from handling a toxic material might be associated with accidental ingestion (contamination of food or drink); entry of particles into the bloodstream (via skin or a cut) or entry of particles into the lung (breathing dust).

The toxic effects on rats from ingestion, subcutaneous implantation and intraperitoneal insertion of a number of different TL phosphors has been studied (Dettmer and Galkin 1968). Their results showed that of the phosphors ingested (i.e. LiF, $Li_2B_4O_7$, CaF_2 and BeO), only LiF had a significant acute effect on the health of the animals. These animals displayed some of the symptoms of fluoride poisoning, i.e. anorexia, weight loss, irritability, lethargy and tremors. This is not surprising, since preparations containing sodium fluoride have been used for many years as rat poisons. Many of the cases of fluoride poisoning in humans have resulted from accidental ingestion of rat poisons containing sodium fluoride (Polson and Tattersall 1959). However, the solubility of sodium fluoride is about ten times that of lithium fluoride. The probable lethal dose in humans for all fluoride salts (with the exception of almost insoluble calcium fluoride) is between 50 and 500 mg kg^{-1} body weight (Gleason *et al* 1969).

Subcutaneous exposure to LiF may cause reversible or irreversible changes to exposed tissue, but not permanent damage according to Sax

(1979). This confirms the observation of Dettmer and Galkin that there were local tissue reactions but no gross effects. However, difficulties were experienced in recovering extruded LiF dosemeters which dissolved in the body tissue.

BeO powder is extremely toxic if inhaled, and prolonged exposure to BeO dust causes chemical pneumonia. It may also act locally on the skin. Implantation and subcutaneous studies on experimental animals show that BeO can cause toxic effects and form neoplasms in doses greater than 10 mg kg^{-1} body weight (NIOSH 1976).

Conclusions

(i) There do not appear to be any published human toxicity data for LiF. Fluoride salts in general are very toxic (with the exception of CaF_2) and NaF, which has a much higher solubility than LiF, can kill (between 3 and 30 g for a 70 kg man). Experimental animals which have ingested LiF dissolved in water show symptoms of fluoride poisoning and may die.

(ii) While Dettmer and Galkin could produce no acute ill effects in rats due to the ingestion of lithium borate, caution should be exercised as there is evidence that it is toxic in man. Boric salts have been used for many years as antiseptics, and at one time boric acid was added to food as a preservative (a practice long since discontinued). Cases of poisoning have occurred as a result of both accidental ingestion and topical application.

(iii) The normal laboratory rules of good housekeeping should ensure that food and drink do not become contaminated with phosphor powder. Under these conditions LiF and $Li_2B_4O_7$ can be regarded as safe.

(iv) While subcutaneous insertion and implantation of LiF do not appear to have any permanent injurious effects, the performance of the dosemeters may be affected greatly by contact with body fluids. They should be sealed in a protective envelope, such as catheter tubing.

(v) BeO powder is extremely toxic if breathed in. The normally used ceramic forms of BeO are very stable and should not present a hazard if carefully handled. BeO should not be directly implanted in tissue but should be contained within a protective envelope.

References

Adam G and Katriel J 1971 *Proc. 3rd Int. Conf. on Luminescence Dosimetry, Riso, Denmark* ed V Mejdahl (Riso: Danish Atomic Energy Research Establishment) *Riso Rep. No. 249* p9

Aitken M J 1968 *Proc. 2nd Int. Conf. on Luminescence Dosimetry, Gatlinburg,* ed J A Auxier *et al* (Tennessee: USAEC and Oak Ridge National Laboratory) *Conf-680920* p281

Aitken M J and Murray A S 1976 *Symp. on Archaeometry and Archaeological Prospection* ed H McKerell (London: HMSO).

Almond P R and McCray K 1970 *Phys. Med. Biol.* **15** 335

ANSI 1975 *Performance, Testing and Procedural Specification for Thermoluminescence Dosimetry (Environmental Applications)* (New York: American National Standards Institute) *ANSI N545–1975*

Attix F H 1974 *Proc. 4th Int. Conf. on Luminescence Dosimetry, Krakow* ed T Niewiadomski (Krakow: Institute of Nuclear Physics) p31

——1975 *J. Appl. Phys.* **46** 81

Ayyangar K, Lakshmanan A R, Bhuwan C and Ramadas K 1974 *Phys. Med. Biol.* **19** 665

Ayyangar K, Reddy A R and Brownell G L 1968 *Proc. 2nd Int. Conf. on Luminescence Dosimetry, Gatlinburg* ed J A Auxier *et al* (Tennessee: USAEC and Oak Ridge National Laboratory) *Conf-680920* p525

Bailiff I K, Bowman S E G, Mobbs S F and Aitken M J 1977 *J. Electrostatics* **3** 269

Bartlett D T 1979 private communication

Bartlett D T and Edwards A A 1979 *Phys. Med. Biol.* **24** 1276

Bartlett D T, McKinlay A F and Smith P A 1980 *Nucl. Technol.*

Bassi P, Busuoli G and Rimondi O 1976 *Int. J. Appl. Rad. and Isotopes* **27** 291

Becker K, Tham T D and Haywood F F 1973 *Proc. 3rd Int. Congr. Int. Radiation Protection Ass.* (Oak Ridge, Tennessee: USAEC) *Conf-730907-P1* p584

Benko L, Uchrin Gy and Biro T 1977 *Proc. 4th Int. Congr. Int. Radiation Protection Ass.* (Paris: International Radiation Protection Association) p1261

Bhasin B D, Sasidharan R and Sunta C M 1976 *Health Phys.* **30** 139

Binder W, Disterhoft S and Cameron J R 1968 *Proc. 2nd Int. Conf. on Luminescence Dosimetry, Gatlinburg* ed J A Auxier *et al* (Tennessee: USAEC and Oak Ridge National Laboratory) *Conf-680920* p43

Bistrović M, Maricić Z, Greenfield M A, Breyer B, Dvornik I, Slaus I and Tomas P 1976 *Phys. Med. Biol.* **21** 414

Bjarngard B 1965 *Proc. Int. Conf. on Luminescence Dosimetry, Stanford* ed F H Attix *et al* (Springfield: USAEC) p195

Bloch P 1968 *Proc. 2nd Int. Conf. on Luminescence Dosimetry, Gatlinburg* ed J A Auxier *et al* (Tennessee: USAEC and Oak Ridge National Laboratory) *Conf-680920* p317

Blum E, Bewley D K and Heather J D 1972 *Phys. Med. Biol.* **17** 661

——1973 *Phys. Med. Biol.* **18** 226

Booth L F, Johnson T L and Attix F H 1972 *Health Phys.* **23** 137

Böttor-Jensen L 1970 *Advances in Physical and Biological Radiation Detectors, Proc. IAEA Symp. Vienna* (Vienna: IAEA) *STI/PUB/269* p20

Böttor-Jensen L and Christensen P 1972 *Acta. Radiol. Suppl.* **313** 247

Boyle R 1663 (1964) *Experiments and Considerations Touching Colours* p413

Bradbury M H and Lilley E 1977 *J. Phys. D: Appl. Phys.* **10** 1261

British Journal of Radiology 1978 *Suppl. No. 11: Central Axis Depth Dose Data for Use in Radiotherapy* ed M Cohen *et al* (London: British Institute of Radiology).

Brunskill R T 1968 *The Preparation and Properties of Thermoluminescent Lithium Borate UKAEA Prog. Rep. 837(W)* (London: UKAEA)

Burgkhardt B, Herrera R and Piesch E 1977 *Proc. 5th Int. Conf. on Luminescence Dosimetry, Sao Paolo* ed A Scharmann (Giessen: Justus Liebig Universitat) p75

Burlin T E 1966 *Br. J. Radiol.* **39** 727

Busuoli G, Cavallini A, Fasso A and Rimondi O 1970 *Phys. Med. Biol.* **15** 673

Busuoli G, Sermenghi I, Rimondi O and Vicini G 1977 *Nucl. Instrum. Meth.* **140** 385

Cameron J R, Daniels F, Johnson N M and Kenney G 1961 *Science* **134** 233

Cameron J R, Zimmerman D W and Bland R 1965 *Proc. Int. Conf. on Luminescence Dosimetry, Stanford* ed F H Attix *et al* (Springfield: USAEC) p47

Cameron J R, Suntharalingam N and Kenny G N 1968 *Thermoluminescent Dosimetry* (Madison, USA: University of Wisconsin Press) p182

Charles M W 1977 *Proc. 5th Int. Conf. on Thermoluminescence Dosimetry, Sao Paolo* ed A Scharmann (Giessen: Justus Liebig Universitat) p313

Chen R and Winer S A A 1970 *J. Appl. Phys.* **41** 5227

Christensen P 1967 *Manganese-activated Lithium Borate as a Thermoluminescent Dosimetry Material* (Riso: Danish Atomic Energy Commission Research Establishment) *Riso Rep. No. 161*

——1968 *Proc. 2nd Int. Conf. on Luminescence Dosimetry, Gatlinburg* ed J A Auxier *et al* (Tennessee: USAEC and Oak Ridge National Laboratory) *Conf-680920* p90

Christensen P and Mäjborn B 1980 *Proc. 6th Int. Conf. on Solid State Dosimetry, Toulouse, Nucl. Instrum. Meth.*

Claffy E W, Klick C C and Attix F H 1968 *Proc. 2nd Int. Conf. on Luminescence Dosimetry, Gatlinburg* ed J A Auxier *et al* (Tennessee: USAEC and Oak Ridge National Laboratory) *Conf-680920* p302

CEC Commission of the European Communities 1975a *Technical Recommendations for Monitoring the Exposure of Individuals to External Radiation* (Luxembourg: CEC) *EUR 5287*

——1975b *Technical Recommendations for the Use of TL for Dosimetry* (Luxembourg: CEC) *EUR 5358*

Cox F M 1968 *Proc. 2nd Int. Conf. on Luminescence Dosimetry, Gatlinburg* ed J A Auxier *et al* (Tennessee: USAEC and Oak Ridge National Laboratory) *Conf-680920* p61

Cox F M, Lucas A C and Kapsar B M 1976 *Health Phys.* **30** 135

Crase K W and Gammage R B 1975 *Health Phys.* **29** 739

Crittenden G C, Townsend P D, Gilkes J and Wintersgill M C 1974 *J. Phys. D: Appl. Phys.* **7** 2415

Curie M 1904 (1961) *Radioactive Substances* (New York: Philosophical Library)

Daniels F, Boyd C A and Saunders D F 1953 *Science* **117** 343

Dekker H 1976 *Health Phys.* **30** 399

Dennis J A, Marshall T O and Shaw K B 1974 *The NRPB Automated TLD and Dose*

Record Keeping System (Harwell: National Radiological Protection Board) *NRPB-R32*

Dettmer C M and Galkin B M 1968 *Proc. 2nd Int. Conf. on Luminescence Dosimetry, Gatlinburg* ed J A Auxier *et al* (Tennessee: USAEC and Oak Ridge National Laboratory) *Conf-680920* p944

DHEW 1977 *A Review of the Use of Ionising Radiation for the Treatment of Benign Diseases* vol 1 (Rockville, USA: Bureau of Radiological Health) *HEW Publ. (FDA) 78-8043*

Driscoll C M H 1977 *An Intercomparison of the Thermoluminescent Efficiency of Various Preparations of Lithium Fluoride* (Harwell: National Radiological Protection Board) *NRPB-R68*

——1978 *Phys. Med. Biol.* **23** 777

Duftschmid K E 1980 *Proc. 6th Int. Conf. on Solid State Dosimetry, Toulouse, Nucl. Instrum. Meth.*

Eisenlohr H H and Jayaraman S 1977 *Phys. Med. Biol.* **22** 18

EMI 1979 *Photomultipliers* (trade catalogue)

Engelke M J and Israel H I 1974 *Health Phys.* **27** 173

Fleming S J 1970 *Archaeometry* **12** 133

——1973 *Archaeometry* **15** 13

Fowler J F and Attix F H 1966 *Radiation Dosimetry* vol II 2nd edn, ed F H Attix and W C Roesch (New York: Academic Press)

Furuta Y and Tanaka S 1972 *Nucl. Instrum. Meth.* **104** 365

Garlick G F J 1949 *Luminescent Materials* (Oxford: Clarendon)

Ginther R J and Kirk R D 1957 *J. Electrochem. Soc.* **104** 365

Gleason M N, Gosselin R E, Hodge H C and Smith R P 1969 *Clinical Toxicology of Commercial Products* (Baltimore, USA: Williams and Wilkins)

Gorbics S G 1965 *Proc. Int. Conf. on Luminescence Dosimetry, Stanford* ed F H Attix *et al* (Springfield: USAEC) p167

Graham C L and Homann S G 1978 *Symp. on National and International Standardization of Radiation Dosimetry, Atlanta* (Vienna: IAEA) p335

Griffith R V, Hankins D E, Gammage R B and Tommasino L 1979 *Health Phys.* **36** 235

Grogan D, Bradley R P and Ashmore J P 1980 *Proc. 6th Int. Conf. on Solid State Dosimetry, Toulouse, Nucl. Instrum. Meth.*

Harvey J R and Townsend S 1971 *Proc. 3rd Int. Conf. on Luminescence Dosimetry, Riso, Denmark* ed V Mejdahl (Riso: Danish Atomic Energy Research Establishment) *Riso Rep. No. 249* p1015

Hashizume T, Kato Y, Nakajima T, Toryu T, Sakamoto H, Kotera N and Eguchi S 1971 *Proc. Symp. on Advances in Radiation Detectors, Vienna* (Vienna: IAEA) *IAEA-SM-143/11* p91

Henson A M and Thomas R H 1978 *Health Phys.* **34** 389

Horowitz Y S, Fraier I, Kalefezra J, Pinto H and Goldbart Z 1979a *Phys. Med. Biol.* **24** 1268

Horowitz Y S and Freeman S 1978 *Nucl. Instrum. Meth.* **157** 393

Horowitz Y S, Freeman S and Dubi A 1979b *Nucl. Instrum. Meth.* **160** 317

Horsley R J and Peters V G 1976 *Br. J. Radiol.* **49** 810

IAEA 1979 *Calibration of Dose Meters used in Radiotherapy* (Vienna: IAEA) *Tech. Rep. Ser. No. 185*

ICRP 1977 *Recommendations of the Int. Comm. on Radiological Protection* (Oxford: Pergamon) *ICRP Publ. 26*

ICRU 1980 *Radiation Quantities and Units, Rep. No. 33* (Washington, USA: ICRU)

——1976 *Determination of Absorbed Dose in a Patient Irradiated by Beams of X or Gamma Rays in Radiotherapy Procedures* (Washington, USA: ICRU) *Rep. No. 24*

ISO 1979 *Personal and Environmental Thermoluminescent Dosemeters* 4th draft *ISO/TC85/SC2/WG7*

Jahnert B 1972 *Health Phys.* **23** 112

Jayachandran C A 1970 *Phys. Med. Biol.* **15** 325

Joelsson I and Backström A 1970 *Acta. Radiol. Ther. Phys. Biol.* **9** 233

Joelsson I, Rudén B I, Costa A, Dutreix A and Rosenwald J C 1972 *Acta. Radiol. Ther. Phys. Biol.* **11** 289

Johansson J M, Lindskoug B A A and Nystrom C E 1969 *Acta. Radiol. Ther. Phys. Biol.* **8** 360

Julius M W 1976 *Proc. 9 Jahrestagung Fachverlsand für Strahlenschutz, Alpbach, FS-75-12-T* p125

Julius H W, Verhoef C W, Busscher F and Oterman F 1974 *Proc. 4th Int. Conf. on Luminescence Dosimetry, Krakow* ed T Niewiadomski (Krakow: Institute of Nuclear Physics) p675.

Jun J S and Becker K 1975 *Health Phys.* **28** 459

Karzmark C J, Fowler J F and White J T 1965 *Proc. Int. Conf. on Luminescence Dosimetry, Stanford* ed F H Attix *et al* (Springfield: USAEC) p265

King S D 1974 *Proc. 4th Int. Conf. on Luminescence Dosimetry, Krakow* ed T Niewiadomski (Krakow: Institute of Nuclear Physics) p873

Kirk R D, Schulman J H, West E J and Nash A E 1967 *Proc. Symp. Solid-state Chem. Rad. Dosimetry* (Vienna: IAEA) p91

Koczynski A, Wolska-Witer M, Böttor-Jensen L and Christensen P 1974 *Proc. 4th Int. Conf. on Luminescence Dosimetry, Krakow* ed T Niewiadomski (Krakow: Institute of Nuclear Physics) p641

Kramer R Regulla D F and Drexler G 1977 *Proc. 5th Int. Conf. on Luminescence Dosimetry, Sao Paulo* ed A Scharmann (Giessen: Justus Liebig Universitat) p298

Krasnaya A R, Nosenko G M, Revzin L S and Yasolko V 1961 *Atomnayia Energyia* **10** 630

Lakshmanan A R and Ayyangar K 1976 *Health Phys.* **31** 284

Lakshmanan A R, Rajendram K V, Ayyangar K and Madhvanath V 1976 *Health Phys.* **30** 489

Lakshmanan A R and Vohra K G 1979 *Nucl. Instrum. Meth.* **159** 585

Langmead W A and Wall B F 1976 *Phys. Med. Biol.* **21** 39

Langmead W A, Wall B F and Palmer K E 1976 *Br. J. Radiol.* **49** 956

Liden K 1948 *Acta. Radiol.* **30** 64

Linsley G S and Mason E W 1971 *Phys. Med. Biol.* **16** 695

Lippert J and Mejdahl V 1965 *Proc. Int. Conf. on Luminescence Dosimetry, Stanford* ed F H Attix *et al* (Springfield: USAEC) p195

Lucas A C and Kapsar B M 1977 *Proc. 5th Int. Conf. on Luminescence Dosimetry, Sao Paulo* ed A Scharmann (Giessen: Justus Liebig Universitat) p131

Lucas A C and Rainbolt C 1968 *Proc. 2nd Int. Conf. on Luminescence Dosimetry, Gatlinburg* ed J A Auxier *et al* (Tennessee: USAEC and Oak Ridge National Laboratory) *Conf-680920* p456

McDougall R S and Rudin S 1970 *Health Phys.* **19** 281

Mäjborn B, Böttor-Jensen L and Christensen P 1977 *Proc. 5th Int. Conf. on Luminescence Dosimetry, Sao Paulo* ed A Scharmann (Giessen: Justus Liebig Universitat) p124

Mandeville C E and Albrecht H O 1954 *Phys. Rev.* **94** 494

Marshall M, Douglas J A, Budd T and Churchill W C 1977 *Proc. 4th Int. Congr. Int. Radiation Protection Ass. Paris* (Paris: IRPA) *Commun. No. N307* p1257

Marshall T O, Shaw K B and Mason E W 1971 *Proc. 3rd Int. Conf. on Luminescence Dosimetry, Riso, Denmark* ed V Mejdahl (Riso: Danish Atomic Energy Establishment) *Riso Rep. No. 249* p530

Mason E W 1970 *Phys. Med. Biol.* **15** 79

Mason E W, McKinlay A F and Clark I 1976 *Phys. Med. Biol.* **21** 60

Mason E W, McKinlay A F, Clark I and Saunders D 1974 *Proc. 4th Int. Conf. on Luminescence Dosimetry, Krakow* ed T Niewiadomski (Krakow: Institute of Nuclear Physics) p219

——1975 *Thermoluminescence Sensitivity Variations in LiF:PTFE Dosemeters Incurred by Improper Handling Procedures* (Harwell: National Radiological Protection Board) *NRPB-R37*

Mason E W, McKinlay A F and Saunders D 1977 *Phys. Med. Biol.* **22** 29

Mayhugh M R 1970 *J. Appl. Phys.* **41** 4776

Mayhugh M R, Christy R W and Johnson N M 1968 *Proc. 2nd Int. Conf. on Luminescence Dosimetry, Gatlinburg* ed J A Auxier *et al* (Tennessee: USAEC and Oak Ridge National Laboratory) *Conf-680920* p294

Mehta S K and Sengupta S 1976 *Phys. Med. Biol.* **21** 955

Moore L E 1957 *J. Phys. Chem.* **61** 636

Morgan T J and Bateman L 1977 *Health Phys.* **33** 339

Nakajima T 1972 *Health Phys.* **23** 133

Nakajima T, Kato Y, Kotera N and Eguchi S 1971 *Health Phys.* **21** 118

Nash A E, Attix F H and Schulman J H 1965 *Proc. Int. Conf. on Luminescence Dosimetry, Stanford* ed F H Attix *et al* (Springfield: USAEC) p244

Niewiadomski T 1976 *Confrontation of Thermoluminescence Models in Lithium Fluoride with Experimental Data* (Krakow: Institute of Nuclear Physics) *Health Physics Lab. Rep. 936/D.*

Nink R and Kos H J 1976 *Phys. Stat. Solidi.* (*a*) **35** 121

NIOSH 1976 *Suspected Carcinogens* 2nd edn (Cincinnati, USA: National Institute for Occupational Safety and Health) *HEW Publ. NIOSH 77–149*

Paliwal B R and Almond P R 1975 *Phys. Med. Biol.* **20** 547

Perry K E 1968 *Atomic Energy Establishment Winfrith, Rep. No. 607* (Winfrith: UKAEA)

Pick H 1972 *Optical Properties of Solids* ed F Abeles (Amsterdam: North-Holland)

Piesch E 1980 *Application of TLD systems for Environmental Monitoring, Proc. Appl. TLD Course, Ispra, Italy* (Bristol: Adam Hilger)

Piesch E, Burgkhardt B and Sayed A M 1978 *Nucl. Instrum. Meth.* **157** 179

Polson C J and Tattersall R N 1959 *Clinical Toxicology* (London: English Universities Press)

Portal G, Berman F, Blanchard P and Prigert R 1971 *Proc. 3rd Int. Conf. on Luminescence Dosimetry, Riso, Denmark* ed V Mejdahl (Riso: Danish Atomic Research Energy Establishment) *Riso Rep. No. 249* p410

Prokic M 1980 *Proc. 6th Int. Conf. on Solid State Dosimetry, Toulouse, Nucl. Instrum. Meth.*

Puite K J and Crebolder D L J M 1974 *Phys. Med. Biol.* **19** 341

Randall J F and Wilkins M H F 1945 *Proc. R. Soc.* **A184** 366, 390

Regulla D F 1971 *Experiences with the LiF TLD System and Recommendations for its Practical Application* (Neuherberg: Gesellschaft für Strahlen und Umweltforschung) *GSF Rep. S-124*

Rieke J R and Daniels F 1957 *J. Phys. Chem.* **61** 629

Robertson M E A 1975 *Identification and Reduction of Errors in TLD Systems* (Weybridge: D A Pitman)

Rossiter M J, Rees-Evans D B, Ellis S C and Griffiths J M 1971 *J. Phys. D: Appl. Phys.* **4** 1245

Rudén B I 1976 *Acta Radiol. Ther. Phys. Biol.* **15** 447

Rudén B I and Bengtsson L G 1974 *Phys. Med. Biol.* **19** 186

——1977 *Acta Radiol. Ther. Phys. Biol.* **16** 157

Rzyski B and Nambi K S V 1977 *Proc. 5th Int. Conf. on Luminescence Dosimetry, Sao Paulo* ed A Scharmann (Giessen: Justus Liebig Universitat) p197

Sax N I 1979 *Dangerous Properties of Industrial Materials* 5th edn (New York: Van Nostrand Reinhold)

Scarpa G 1970a *Health Phys.* **19** 91

——1970b *Phys. Med. Biol.* **15** 667

Scarpa G, Moscati M and Furetta C 1979 *Br. J. Radiol.* **52** 75

Schayes R, Brooke C, Kozlowitz I and L'Hereux M 1965 *Proc. Int. Conf. on Luminescence Dosimetry, Stanford* ed F H Attix *et al* (Springfield: USAEC) p138

Schulman J H, Attix F H, West E J and Ginther R J 1960 *Rev. Sci. Instrum.* **31** 1263

Schulman J H, Kirk R D and West E J 1965 *Proc. Int. Conf. on Luminescence Dosimetry, Stanford* ed F H Attix *et al* (Springfield: USAEC) p113

Sievert R M 1934 *Acta Radiol.* **15** 193 (in German)

Spanne P 1974 *Proc. 4th Int. Conf. on Luminescence Dosimetry, Krakow* ed T Niewiadomski (Krakow: Institute of Nuclear Physics) p597

——1979 *Acta Radiol.* Suppl. **360**

Spurny Z and Hruska J 1976 *Phys. Med. Biol.* **21** 439

Stoebe T G and Watanabe S 1975 *Phys. Stat. Solidi (a)* **29** 11

Storm E and Israel H I 1967 *Photon Cross Sections from 0.001 to 100 MeV for Elements 1 through 100* (Los Alamos: Clearing House of Scientific and Technical Information) *Rep. No. 23753*

Suntharalingam N and Cameron J R 1969 *Phys. Med. Biol.* **14** 394

Suntharalingam N and Mansfield C M 1971 *Proc. 3rd Int. Conf. on Luminescence Dosimetry, Riso, Denmark* ed V Mejdahl (Riso: Danish Atomic Energy Research Establishment) *Riso Rep. No. 249* p816

Sutton S R and Zimmerman D W 1976 *Archaeometry* **18** 125

Tanaka S and Furuta Y 1974 *Nucl. Instrum. Meth.* **117** 93

Takenaga M, Yamamoto O and Yamashita T 1977 *Proc. 5th Int. Conf. on Luminescence Dosimetry, Sao Paulo* ed A Scharmann (Giessen: Justus Liebig Universitat) p148

——1977 *Nucl. Instrum. Meth.* **140** 395

Taylor F E and Webb G A M 1978 *Radiation Exposure of the UK Population* (Harwell: National Radiological Protection Board) *NRPB-R77*

Thompson J J and Ziemer P L 1973 *Health Phys.* **25** 435

Tochilin E and Goldstein N 1966 *Health Phys.* **12** 705

Tochilin E, Goldstein N and Lyman J F 1968 *Proc. 2nd Int. Conf. on Luminescence*

Dosimetry, Gatlinburg ed J A Auxier *et al* (Tennessee: USAEC and Oak Ridge National Laboratory) *Conf-680920* p424

Tochilin E, Goldstein N and Miller W G 1969 *Health Phys.* **16** 1

Turner A F and Anderson D W 1973 *Phys. Med. Biol.* **18** 46

Vacirca S J, Thompson D L, Pasternack B S and Blatz H 1972 *Phys. Med. Biol.* **17** 71

Vora H, Jones J H and Stoebe T G 1975 *J. Appl. Phys.* **46** 71

Wald J, Dewerd L A and Stoebe T G 1977 *Health Phys.* **33** 303

Wall B F, Fisher E S, Paynter R, Hudson A and Bird P D 1979a *Br. J. Radiol.* **52** 727

Wall B F, Green D A C and Veerappan R 1979b *Br. J. Radiol.* **52** 189

Wallace R H and Ziemer P L 1968 *Proc. 2nd Int. Conf. on Luminescence Dosimetry, Gatlinburg* ed J A Auxier *et al* (Tennessee: USAEC and Oak Ridge National Laboratory) *Conf-680920* p140

Walton A J 1977 *Adv. Phys.* **26** 887

Watanabe K 1951 *Phys. Rev.* **83** 785

Wilson C R, Dewerd L A and Cameron J R 1966 *USAEC Rep. C00-1105-116*

Wingate C L, Tochilin E and Goldstein N 1965 *Proc. Int. Conf. on Luminescence Dosimetry, Stanford* ed F H Attix *et al* (Springfield: USAEC) p421

Yamaoka Y 1978 *Health Phys.* **35** 708

Yamashita T, Nada N, Onishi H and Kitamura S 1971 *Health Phys.* **21** 295

Zanelli G D 1968 *Phys. Med. Biol.* **13** 393

Zimmerman D W 1971 *Archaeometry* **13** 29

Zimmerman D W, Rhyner C R and Cameron J R 1965 *Proc. Int. Conf. on Luminescence Dosimetry, Stanford* ed F H Attix *et al* (Springfield: USAEC) p86

Zimmerman J 1971 *J. Phys. C: Solid St. Phys.* **4** 3277

Index

LIVERPOOL
UNIVERSITY
FIAT LVX